DATE DUE

AUG 1 3 1996			
AUG 0 2 1996			
JAN 3 0 2009	FEB 1 3		

CARR McLEAN, TORONTO FORM #38-297

No. 1233
$15.95

AC/DC Electricity And Electronics Made Easy

by Victor F. Veley

162668316

 **TAB BOOKS Inc.**
BLUE RIDGE SUMMIT, PA. 17214

FIRST EDITION

SECOND PRINTING

Library of Congress Cataloging in Publication Data

Veley, Victor F. C.
    AC/DC electricity and electronics made easy.

    Includes index.
    1. Electricity.   2. Electronics.   I. Title.
AC522.V44   537   80-28271
ISBN 0-8306-9651-2
ISBN 0-8306-1233-5 (pbk.)

Preface

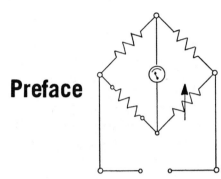

Technicians and students of electronics technology are often required to substitute numerical values in an expression to derive a particular answer. There are only two problems. First of all, it is necessary to remember the formula correctly and this may not be too easy, especially if there are various powers of ten and constants involved. Second, one must recognize the units in which the answer is obtained and then convert, if required, to practical values. This is certainly true if one is concerned with electromagnetic units (emu) or electrostatic units (esu).

How much simpler it would be if there were only one system of electrical units! In 1901, Professor Giorgi of Italy proposed a system based on the meter, kilogram and second for length, mass and time with the choice of a fourth fundamental unit being the ampere. Since 1950, more and more countries have officially recognized this system and there is little doubt that, in the future, it will be universally adopted. The advantages lie in the simpler formulas which always employ these practical units and the elimination of the confusion existing between the previous systems. This was achieved at the expense of introducing certain additional units and assigning new values to the permeability and permittivity of free space.

I had taught electronics for many years using the old methods and was agreeably surprised to find how much simpler the mks system was for the instructor to use and for the student to understand. This book is therefore the outcome of my experience in teaching to a variety of classes the subjects of direct and alternating current.

The purpose of this book is twofold. First it should prove to be a valuable reference for the technician in the field. Each chapter is organized into a number of sections. In each section one or more important concepts are introduced with the relevant equations. The concepts and the equations are then used in a number of comprehensive examples which are accompanied by detailed solutions. Finally, at the end of each chapter, is a summary of the more important equations and, where necessary, a table is included to illustrate the differences between the modern SI units and the older cgs systems. The technician is therefore given every opportunity to familiarize himself with the modern system of units.

The second purpose is to assist those community college students in the first and second semesters of electronics technology. The survey of SI units in this reference manual will be a valuable supplement to any standard textbook which the student is using. In addition, the examples chosen are unusually comprehensive and should provide the student with a better understanding of the principles involved. In virtually every example only standard practical values are used, and many problems are solved by a variety of methods. This serves as an accurate check on the answers obtained and as a comparison between various analytical techniques. Detailed appendices are included on the subjects of decibels and waveform analysis.

I wish to acknowledge a great debt to my wife, Joyce, for typing the manuscript and for making this work possible.

Victor F. Veley

Contents

Mechanical Units

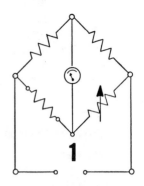

A system of units represents the language in which a scientific subject is discussed. All such systems are built on certain fundamental units which are taken as standards.

FUNDAMENTAL UNITS OF THE MKS SYSTEM

In the *mks system* the units of mass is the kilogram (1,000 grams) which is defined by an artificial standard held by an international bureau in France. The unit of length is the *meter* which is derived from an atomic standard, while the second is $\frac{1}{86,400}$ of the mean solar day.

UNIT OF FORCE

When a force is applied to a mass, the mass accelerates (resulting in a change in velocity). In the mks system, acceleration is measured in meters per second per second. For example, if a mass starts from rest (zero velocity) with a constant acceleration of 2 meters per second per second, its velocity after 1 second is 2 meters per second. After 2 seconds it is 4 meters per second and so on. There are many units of force in existence but in the mks system, a new one is introduced. This is the *newton* (named after Issac Newton, 1642 - 1727) which is the force required to give a mass of 1 kilogram an acceleration of 1 meter per second per second in the direction of the force. F(force in newtons) = M(mass in kilograms) X A(acceleration in meters per second per second).

The force of gravity gives a mass an acceleration of approximately 9.81 meters per second per second. This is not constant but varies slightly over different positions on the earth's surface. The force exerted on a mass of 1 kilogram due to gravity = 9.81 newtons; this force is sometimes referred to as a 1 kilogram weight. On the moon, the gravitational force on 1 kilogram would be much less than 9.81 newtons.

In the old centimeter-gram-second system, the unit of force was the *dyne*, which gave 1 gram an acceleration of 1 centimeter per second per second. Since 1 kilogram = 1000 gms and 1 meter = 100 centimeters, 1 newton = 100,000 dynes. For most purposes the dyne is far too small and therefore the newton has a practical advantage.

UNIT OF ENERGY OR WORK

When a force (newtons) is exerted through a distance (meters) in the direction of the force, work is done or energy is consumed (energy is the capacity for doing work). In the mks system the unit of work is the *joule* (named after James Joule, 1518 - 89) which is equivalent to 1 meter-newton.

Work done = F(force in newtons) × D(distance in meters) = F × D (joules or meter-newtons) (*not* newtons – meters.) See unit of torque.

Since the various forms of energy are interchangeable, the joule will also be the mks unit of electrical energy. In the cgs system, the *erg* is the work done when a force of 1 *dyne* acts through 1 centimeter in the direction of the force. 1 joule = 1 meter-newton = 100 × 100000 = 10,000,000 ergs.

The unit of heat energy is the *calorie* which is the amount of heat required to raise the temperature of 1 gram of water by 1° Centigrade. Experimentally it has been found that 1 calorie = 4.18 joules.

UNIT OF TORQUE

Work is the product of force and distance but to obtain *torque*, it is also necessary to multiply these quantities together. However, for work, the force and distance lie in the same direction, whereas for torque, the force and the distance are at right angles to each other. In Fig. 1-1, 0 is a pivotal point and a force F is acting at a point P, distance D from O. The torque of F about O = F × D newton-meters (*not* joules).

In cgs units, torque is measured in dyne-centimeters.

POWER

This is the *rate* at which work is done or energy is produced. It is sometimes difficult to distinguish between power and work. For example, a child may use a complex system of pulleys to lift a heavy weight against gravity while a strong man may swiftly lift the

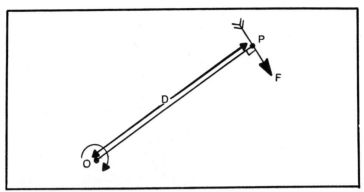

Fig. 1-1. Torque.

weight through the same distance. Neglecting losses in the pulley system, the man and the child performed the same total work but since the man took a shorter time, his power was greater. In the mks System, the unit of power is the *watt* (named after James Watt, 1736 - 1819). It is equal to 1 joule of energy being produced for released per second. A 60 watt electric light bulb will dissipate 60 joules of energy every second (mostly in the form of heat but a small amount as light). Watts = joules per second or joules = watts × seconds.

The unit on your electric bill is a measure of the work done. The joule is too small for this purpose and the unit chosen was the kilowatt-hour (kwh) = 1000 watts × 60 minutes × 60 seconds = 3,600,000 joules.

Another mechanical unit of power is the horsepower which is equal to 746 watts.

PREFIXES

A number of prefixes are used in electronics to create multiples or submultiples of basic units. The most important are:

	Abbreviation	Factor
Tera -	T	× 1,000,000,000,000 or 10^{12}
Giga -	G	× 1,000,000,000 or 10^{9}
Mega -	M	× 1,000,000 or 10^{6}
Kilo -	k	× 1,000 or 10^{3}
Milli-	m	÷ 1,000 or × 10^{-3}
Micro -	μ	÷ 1,000,000 or × 10^{-6}
Nano -	n	÷ 1,000,000,000 or × 10^{-9}
Micro-Micro	$\mu\mu$ or	÷ 1,000,000,000,000 or × 10^{-12}
or Pico -	P	

Example 1-1

A force of 200 newtons is applied to a mass of 50 kg. Find
(a) the acceleration
(b) the velocity after 5 seconds and
(c) the work done during this time and the average power.

Solution

(a) Force = mass × acceleration

200 newtons = 50kg × acceleration

$$\text{Acceleration} = \frac{200 \text{ newtons}}{50 \text{kg}} = 4 \text{ meters per second per second.}$$

(b) Velocity after 5 seconds = Acceleration × time

= 4 meters per second per second × 5 seconds

= *20 meters per second.*

(c) Average velocity = $\frac{20}{2}$ = 10 meters per second

Distance covered = average velocity × time

= 10 meters per second × 5 seconds

= 50 meters

Work done = 200 newtons × 50 meters

= 10,000 joules

Average power = $\frac{10,000 \text{ joules}}{5 \text{ seconds}}$

= 2,000 watts

= *2kw.*

Example 1-2

A metal block whose mass is 180 grams is given an acceleration of 25 cm/sec^2. What is the accelerating force in newtons? What is the kinetic energy in joules when the block's velocity is 450 cm/sec?

Solution

Force = mass (kg) × acceleration (m/sec^2)

$$= \frac{180}{1000} \times \frac{25}{100}$$

= *0.045 newton.*

$$\text{Kinetic energy} = \frac{1}{2} \times \text{mass (kg)} \times (\text{velocity (m/sec)})^2$$

$$= \frac{1}{2} \times \frac{180}{1000} \times \left(\frac{450}{100}\right)^2$$

$$= 1.82 \, joules.$$

Example 1-3

A mass of 1600 kg is lifted vertically with a velocity of 90 meters per minute. Calculate

(a) the power required in kilowatts and

(b) the kinetic energy of the load in joules.

Solution

Force required = 1600 × 9.81 newtons

$$\text{Velocity} = \frac{90}{60} = 1.5 \, \text{meters per second.}$$

$$\text{Power} = 1600 \times 9.81 \times 1.5 \, \text{W} = \frac{1600 \times 9.81 \times 1.5 \text{kW}}{1000}$$

$$= 23.5 \, kW.$$

$$\text{Kinetic energy} = \frac{1}{2} \times 1600 \times (1.5)^2$$

$$= 1800 \, joules.$$

Example 1-4

A motor is developing 60 horse-power with a speed of 800 rpm. Calculate its torque in newton-meters.

Solution

60 hp = 60 × 746 W = 44760 W.

$$\text{Angular velocity} = 2\pi \times \frac{800}{60} = 83.776 \, \text{radians per second.}$$

$$\text{Torque} = \frac{44760}{83.776} = 534 \, newton\text{-}meters.$$

Example 1-5

A mass of 175 kg is pulled along a horizontal surface whose coefficient of friction is 0.32. If the mass covers a distance of 25 meters in 5.7 seconds, calculate

    (a) the horizontal force required in newtons,

    (b) the work done in joules and

    (c) the work power in watts.

Solution

    (a) Horizontal force $= 175 \times 9.81 \times 0.32 = 552.5 \; newtons$.

    (b) Work done $= 552.5 \times 25 = 13812 \; joules$.

    (c) Power $= 13812/5.7 = 2423 \; watts$.

CHAPTER SUMMARY

☐ F(force in newtons) = M(Mass in kilograms) × A(acceleration in meters per second per second)

☐ 1 newton = 100,000 dynes

☐ Work done (joules) = F(force in newtons) × D(distance in meters)

☐ 1 joule = 1 watt-second = 10,000,000 ergs

☐ 1 kilowatt-hour (kwh) = 3600000 joules

☐ Torque (newton meters) = F(force in newtons) × D(distance in meters)

☐ 1 calorie = 4.18 joules

☐ Power (watts) = joules per second

☐ 1 horse - power = 746 watts

Electrical Units and Ohm's Law

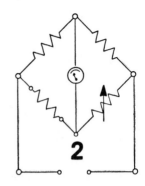

2

In a system of electrical units it is necessary to define a fundamental unit and then build on this foundation to establish other units. In the rationalized mks or S.I. system (Systéme Internationale d´Unité s) the fundamental unit of current internatially adopted in 1948. An electric current is the rate at which a quantity of electricity or charge flows past a given point in an electrical circuit. The unit of current is called the *ampere*, which commemorates a famous French scientist, André -Marie Ampére (1775 - 1836).

UNITS OF CURRENT AND CHARGE

The value of the ampere, commonly abbrevated to amp. or "A", is defined by an effect which occurs in electromagnetism. It is that value of current which, when flowing in each of two indefinitely long parallel conductors, whose centers are separated by 1 meter, in a vacuum, causes a force of 2×10^{-7} newtons per meter length, to be exerted on each conductor. The apparatus which is needed to measure a current accurately from this definition is so very expensive and complex that it is normally not available outside of national laboratories.

The unit quantity of electricity or unit of charge is called the *coulomb*, which is named after another French scientist, Charles A Coulomb. When a current of 1 ampere is maintained for 1 second, a charge of one coulomb flows past a given point in the circuit. It therefore follows that if a current of I amperes is maintained at a constant level for t seconds, the corresponding charge of Q coulombs is given by the following formula. Q (coulombs, abbreviated is C) = I (amperes) × t (seconds)

Consequently

$$I \text{ (amperes)} = \frac{Q \text{ (coulombs)}}{t \text{ (seconds)}}$$

Notice that with the aid of the time unit we have derived the coulomb from the ampere.

A charge of 1 coulomb is equivalent to the total negative charge possessed by 6.24×10^{18} electrons. This means that the charge carried by a single electron is $1/6.24 \times 10^{18} = 1.6019 \times 10^{-19}$ coulomb, while its mass is 9.1066×10^{-31} kilogram.

ELECTROCHEMICAL EQUIVALENT

When a current of 1 ampere is passed between two copper plates immersed in an electrolyte of copper sulphate solution (a copper voltameter), copper is deposited at the rate of 0.0000003294 kilograms per second. The mass of 0.0000003294 kilograms is therefore liberated from the electrolyte by 1 coulomb. In a silver voltameter (two silver plates immersed in a silver nitrate solution), 1 coulomb liberates 0.0000011182 kilograms of silver. The amount of electrolyte liberated by 1 coulomb is called the *electrochemical equivalent*. The electrochemical equivalents of copper and silver are 3.294×10^{-7} kilograms per coulomb (kg/C) and 1.1182×10^{-6} kilograms per coulomb. For nickel and zinc the electrochemical equivalents are respectively 3.04×10^{-7} and 3.38×10^{-7} kg/C.

If z is the electrochemical equivalent of a substance in kilograms per coulomb and I is the current in amperes maintained over an interval of t seconds, then the mass of the substance liberated $\times$ z It kilograms.

Since 1 ampere is 1 coulomb per second, the unit of charge may also be called an ampere-second. A larger unit is the ampere-hour (Ah) which is equivalent to $60 \times 60 = 3600$ coulombs. For example, if a battery delivers a current of 3 amperes for 5 hours, the charge lost is 3A $\times$ 5h = 15 Ah, or $15 \times 3600 = 54,000$ coulombs.

Example 2-1

If a steady current of 12 A is maintained at a point for a time of 5 minutes, calculate the charge flowing past that point in (a) coulombs and (b) ampere-hours.

Solution

Charge = I (amperes) $\times$ t (seconds)

$= 12 \times 5 \times 60$

$= 3600 \text{ coulombs}$

$$=\frac{3600}{3600} \text{ amperes-hour}$$

$$=1 \text{ ampere-hour}.$$

Example 2-2

A charge of 750 coulombs flows past a point in 5 minutes. Calculate the current in amperes.

Solution

Current = Q (coulombs)/t (seconds) = 750/(5×60) = $2.5A$.

Example 23

A constant current of 7 A flows through a copper sulphate solution for a time interval of 50 minutes. Calculate the mass of copper liberated.

Solution

Mass of copper liberated = z It
$$= 3.294 \times 10^{-7} \times 7 \times 50 \times 60$$
$$= 6.92 \times 10^{-3} kilogram.$$

UNIT OF ELECTROMOTIVE FORCE AND POTENTIAL DIFFERENCE

A source of electrical energy such as a battery provides an electromotive force (EMF) which impels electricity through a conductor connected across the battery terminals. The EMF is then balanced by an equal potential difference (p.d) or difference of potential across the ends of the conductor. For both EMF and potential difference the unit is the Volt (V), named after the discoverer of the first electrical battery, the Italian Count Alessandro Volta (1745 - 1827). In Fig. 2-1 E represents the electromotive force and V is the potential difference.

When a charge is driven between two points possessing a potential difference, work must be done. The volt may therefore be defined in term of the coulomb and the joule. There is no need to redefine the joule since it has already been established during our derivation of the mechanical units. The relationship is: Work done (joules) = potential difference (volts) × charge (coulombs) or the

$$\text{Potential Difference (volts)} = \frac{\text{Work done (joules)}}{\text{Charge (coulombs)}}.$$

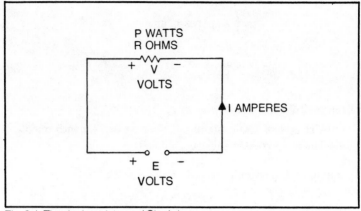

Fig. 2-1. The single resistor and Ohm's law.

In chapter one, we established that

$$\text{Power (Watts)} = \text{joules per second}$$
$$= \text{volts} \times \text{coulombs per second}$$
$$= \text{volts} \times \text{amperes}.$$

Consequently Power, $P = IV$ watts

Current, $I = P/V$ amperes

Voltage, $V = P/I$ volts.

The power in watts is therefore equal to the result of multiplying the current in amperes by the EMF or potential difference in volts.

Example 2-4

A charge of 5 coulombs is moved through a potential difference of 15 volts in a time of 3 seconds. Calculate the work done and the power involved.

Solution

Work done = 15 volts × 5 coulombs

= 75 joules.

Power = 75 joules/3 seconds *= 25 watts.*

Example 2-5

A current of 6 amperes is associated with a potential difference of 75 volts for a period of 3 hours. Calculate the power and the total work in kilowatt-hours.

Solution

Power = 6 amperes × 75 volts.

= 450 watts.

$$\text{Total work} = 450 \text{ watts} \times 3 \text{ hours}$$
$$= 1350 \text{ watt-hours}$$
$$= 1.35 \, kWh.$$

OHM'S LAW AND THE RESISTOR COLOR CODE

In 1827 Dr. George Simon Ohm discovered that, under constant physical conditions, the current through a conductor was directly proportional to the difference of potential across the conductor. In other words, tripling the voltage, tripled the current and halving the voltage, halved the current. Looked at another way, the ratio of the potential difference in volts to the current in amperes is a constant and this relationship is known as *Ohm's Law* of constant proportionality. The constant is a measure of the conductor's opposition to current flow and is called the resistance R. The unit of resistance is the *ohm* and this symbol is the Greek letter Ω (omega).

Referring to Fig. 2-1, Ohm's Law in equation form is: $\dfrac{V}{I} = R$.

This leads to: $V = I \times R$ and $I = \dfrac{V}{R}$.

Also: $P = IV = \dfrac{V}{R} \times V = \dfrac{V^2}{R} = \dfrac{V^2}{R}$ watts or $P = IV = I \times I$

$\times R = I^2 R$ watts.

Summarizing, the twelve relationships involving V, I, R and P are:

$$V = IR = \frac{P}{I} = \sqrt{P \times R} \text{ volts.}$$

$$I = \frac{V}{R} = \frac{P}{V} = \sqrt{\frac{P}{R}} \text{ amps.}$$

$$R = \frac{V}{I} = \frac{E}{I^2} = \sqrt{\frac{P}{I}} \text{ ohms.}$$

$$P = IV = I^2 R = \frac{V^2}{R} \text{ watts.}$$

Table 2-1. EIA Resistor Color Code.

COLOR	BAND 1 (First Significant Figure)	BAND 2 (Second Significant Figure)	BAND3 (Multiplier)	BAND 4 (Tolerance, Percent)
Black	0	0	$\times 10^0$	
Brown	1	1	$\times 10^1$	
Red	2	2	$\times 10^2$	
Orange	3	3	$\times 10^3$	
Yellow	4	4	$\times 10^4$	
Green	5	5	$\times 10^5$	
Blue	6	6	$\times 10^6$	
Violet	7	7	$\times 10^7$	
Gray	8	8	$\times 10^8$	
White	9	9	$\times 10^9$	
None				20
Silver			$\times 10^{-2}$	10
Gold			$\times 10^{-1}$	5

Most resistors used in electronics have their ohmic value and tolerance color-coded on the body of the resistor. The Electronics Industry Association code is shown in Table 2-1.

Referring to Fig. 2-2, bands 1 and 2 indicate the first two significant figures of the rated value. Band 3 is the multiplier and band 4 is the tolerance. For example, a resistor with band 1 red, band 2 violet, band 3 orange, band 4 silver would have a rated value of $27 \times 10^3 = 27000 \ \Omega = 27$ kΩ, with a 10 % tolerance. The permitted limits for this resistor would be $27000 + \dfrac{27000 \times 10}{100} =$

$29700 \ \Omega = 2.97$ kΩ, and $27000 - \dfrac{27000 \times 10}{100} = \quad 24300 \ \Omega \quad =$

2.43 kΩ. The common wattage ratings for such resistors are ⅛, ¼, ½, 1 and 2 watts.

Resistors are only commercially manufactured in certain ohmic values, (see Table 2-2). The values as shown can either be followed by the appropriate number of zeros, or be divided by 10 or 100 when gold or silver respectively appears in the third band.

The tolerance is the reason for using standard resistor values. For example the permitted upper limit of a 220 Ω, 20% resistor is

$220 + \dfrac{220 \times 20}{100} = 264 \ \Omega$. The next highest rated value is 330 Ω

which would have a permitted lower limit of $330 - \dfrac{330 \times 20}{100} = 264 \ \Omega$.

The values are therefore chosen to avoid overlap between the tolerance spreads of resistors with adjacent rated values.

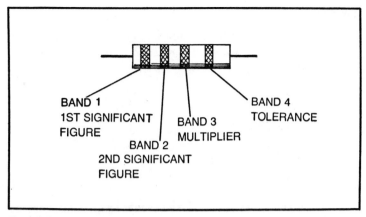

Fig. 2-2. The color coded resistor.

Example 2-6

The first three bands of a color-coded resistor are yellow, violet, orange. Assuming that the resistance is the same as its rated value, what is the magnitude of the voltage applied across the resistor if the current flowing is 435 microamperes?

Table 2-2. Standard Resistor Values.

	5 Percent Tolerance	10 Percent Tolerance	20 Percent Tolerance
	10	10	10
	11		
	12	12	
	13		
	15	15	15
	16		
	18	18	
	20		
	22	22	22
	24		
	27	27	
	30		
	33	33	33
	36		
	39	39	
	43		
	47	47	47
	51		
	56	56	
	62		
	68	68	68
	75		
	82	82	
	91		

Solution

The rated value of the resistor is 4 (yellow), 7 (violet), 000 (orange) Ω, or 47 kΩ = $47 \times 10^3 \Omega$. The current is 435 microamperes $= 435 \times 10^{-6}$ A.

$$E = I \times R$$
$$= 435 \times 10^{-6} \times 47 \times 10^3 \text{ V}$$
$$= 4.35 \times 10^2 \times 10^{-6} \times 4.7 \times 10^1 \times 10^3$$
$$= 4.35 \times 4.7$$
$$= 20.4 \text{ V}.$$

Example 2-7

The current flowing through a particular 1 watt resistor is 23.5 mA. What is the highest value of voltage which can be applied across the resistor without exceeding its power rating?

Solution

Current is $23.5 \text{ mA} = 23.5 \times 10^{-3}$ A.

$$E = \frac{P}{I} = \frac{1}{23.5 \times 10^{-3}} = \frac{1 \times 10^3}{23.5} = \frac{1000}{23.5} = 42.6 \text{ volts.}$$

Example 2-8

The first three bands of a ½ watt color-coded resistor are green, blue, yellow. What is the value of the voltage which can be applied across the resistor without exceeding its power rating?

Solution

Value of the resistor is 5 (green), 6 (blue), 0000 (yellow) Ω = 560 kΩ.

$$E = \sqrt{P \times R} = \sqrt{½ \times 560000}$$
$$= \sqrt{280000}$$
$$= \sqrt{28 \times 10^4} = \sqrt{28} \times 10^2 = 529 \text{ V.}$$

Example 2-9

A 12 V battery is connected across a color-coded resistor whose first three bands are red, red, green. Assuming that the resistance is the same as its rated value, what is the value of the current flowing through the resistor?

Solution

The value of the resistor is 2 (red), 2 (red), 00000 (green) Ω = 2200 kΩ = 2.2 MΩ.

$$I = \frac{E}{R} = \frac{12}{2.2 \times 10^6} = \frac{12 \times 10^{-6}}{2.2} \; A = \mathit{5.5 \, \mu A}.$$

Example 2-10

A 25 watt electric light bulb is operated from a 115 V DC source. What is the value of the current which is flowing in the circuit?

Solution

$$\text{Current, } I = \frac{P}{E} = \frac{25}{115} = 0.217 \, A = \mathit{217} \, mA$$

Example 2-11

The first three bands of a ¼ watt color-code resistor are blue, grey, brown. What is the highest value of current which can flow through the resistor without exceeding its power rating (assuming that the actual resistance is the same as the color-coded value)?

Solution

Value of the resistor is 6 (blue), 8 (grey), 0 (brown) or 680 Ω.

$$\text{Then } I = \sqrt{\frac{P}{R}} = \sqrt{\frac{\frac{1}{4}}{680}} = \sqrt{\frac{0.25}{680}} = \sqrt{0.0003676}$$
$$= \sqrt{3.676 \times 10^{-4}} \, A$$
$$= 1.9 \times 10^{-2} \, A$$
$$= \mathit{19 \, mA}.$$

Example 2-12

When 15.4 V is applied across a resistor, the current is measured as 38 μA. What is the value of the resistor?

Solution

$$\text{Resistance, } R = \frac{E}{I} = \frac{15.4}{38 \times 10^{-6}}$$
$$= \frac{15.4 \times 10^6}{38}$$
$$= 0.405 \times 10^6 \, \Omega = \mathit{405 \, k\Omega}$$

Example 2-13

A 75 watt electric light bulb is designed to operate for a 110 V DC supply. What is the "hot" resistance of the bulb's filament?

Solution

$$\text{Resistance, } R = \frac{E^2}{P} = \frac{110^2}{75} = \frac{110 \times 110}{75} = 161 \ \Omega.$$

Example 2-14

When a current of 237 mA flows through a wire-wound resistor, the power dissipated is 55 W. What is the value of the resistance?

Solution

$$\text{Resistance, } R = \frac{P}{I^2} = \frac{55}{(237 \times 10^{-3})^2}$$

$$= \frac{55}{(2.37 \times 10^{-1})^2}$$

$$= \frac{55}{2.37 \times 2.37 \times 10^{-2}}$$

$$= \frac{55 \times 10^2}{2.37 \times 2.37} = 979 \ \Omega.$$

Example 2-15

When a 220 V DC source is applied across a resistive load, the current taken from the source is 17.3 A. How much power is dissipated in the load?

Solution

$$\text{Power, } P = E \times I = 220 \times 17.3 = 3810 \text{ W} = 3.81 \ kW.$$

Example 2-16

A current of 23.7 mA flows through a 2 watt, 3.3 kΩ resistor. How much power is dissipated in the resistor?

Solution

$$\begin{aligned}\text{Power, } P = I^2 \times R &= (23.7 \times 10^{-3})^2 \times 3.3 \times 10^3 \\ &= (2.37 \times 10^{-2})^2 \times 3.3 \times 10^3 \\ &= 2.37 \times 2.37 \times 3.3 \times 10^{-4} \times 10^3 \\ &= 1.85 \ W.\end{aligned}$$

Example 2-17

The first three bands of a color-coded 2 watt resistor are grey, red, black. If this resistor is connected across a 9 V battery, what is the amount of the power dissipated?

Solution

Value of the resistor is 8 (grey), 2 (red) ohms (the multiplier is black and therefore no zeros appear after the 2). Then

$$P = \frac{E^2}{R} = \frac{9^2}{82} = \frac{81}{82} = 0.99\,W.$$

CONDUCTANCE

The resistance of a conductor is a measure of its opposition to current flow and is defined from the relationship, $R = V/I$. By contrast conductance measures the ease with which current will flow through a conductor; conductance, G, is therefore the inverse

or reciprocal of resistance and is defined by: $\quad G = \dfrac{I}{V} = \dfrac{1}{R}$

On the SI system the unit of conductance is the *siemens* (S). Prior to the introduction of the siemens, the unit of conductance was the *mho* (ohm spelled backwards).

Example 2-18

The first three bands of a color-coded resistor are red, red, gold. What is the resistor's conductance?

Solution

Gold or silver in the third band represent multiplier values of 0.1 and 0.01 respectively. The value of the resistor is therefore 22 × 0.1 = 2.2 Ω. Its conductance is

$$G = \frac{1}{2.2} = 0.45\,S.$$

Example 2-19

The total conductance of a DC circuit is 3.75×10^{-4} S. If the DC source voltage is 165 V, what is the amount of current taken from the source?

Solution

$$\begin{aligned}
I = E \times G &= 165 \times 3.75 \times 10^{-4} \\
&= 1.65 \times 10^2 \times 3.75 \times 10^{-4} \\
&= 6.1875 \times 10^{-2}\,A \\
&= 61.9\,mA.
\end{aligned}$$

Example 2-20

When an EMF of 12 V is applied across a resistive load, the DC load current is 250 mA. What is the conductance of the load?

Solution

$$\text{Conductance},\ G = \frac{I}{E} = \frac{250 \times 10^{-3}}{12} = 0.021\,S.$$

RESISTANCE OF THE CYLINDRICAL CONDUCTOR

The resistance of the cylindrical conductor is directly proportional to the conductor's length, inversely proportional to its cross-sectional area, and depends on the material from which the conductor is made. In equation form $R = \rho \times \dfrac{L}{A}$

R = resistance of the conductor in ohms.
L = length of the conductor in meters.
A = cross-sectional area of the conductor in square meters
ρ (rho) = specific resistance or resistivity of the conductor's material.

With L in meters and A in square meters, the value of ρ will be in SI units. A conductor one meter long with a cross-sectional area of 1 square meter has a resistance to ρ ohms. The SI unit of resistivity is the ohm meter (Ω.m) but since the resistance of a conductor depends on the temperature, the value of the resistivity is quoted for a particular temperature which is normally 20°C (See Table 2-3).

The cgs unit of resistivity is the ohm centimeter (1 ohm meter = 100 ohm centimeters) but in the British system, ρ is measured in ohms circular mil per foot where one circular mil (cmil) is the area of a circle whose diameter is 1 linear mil = 1/1000 inch. The use of circular mils avoids the need for using π when calculating the cross-sectional area. If, for example, the diameter is 3 mils, the area is 3^2 = 9 cmils. In this system of units the resistance of the cylindrical conductor is:

$$R = \frac{L}{A}\rho = \rho\frac{L}{d^2}$$

R = resistance of the conductor in ohms.
ρ = specific resistance in ohm cmil/ft.
L = length of the conductor in feet.
A = cross-sectional area in cmils.
d = diameter in linear mils.

The conversion factor between the SI and British systems is 1 ohm meter = 6.015×10^8 ohm circular mil per foot, so that for annealed copper, $\rho = 1.7 \times 10^{-8}\ \Omega$.m = $1.724 \times 10^{-8} \times 6.015 \times 10^8$ or 10.37 ohm cmil/ft.

Just as conductance is the reciprocal of resistance, conductivity (σ-sigma) is the reciprocal of resistivity. The conductivity is the conductance of a meter length of the material with a cross-

Table 2-3. Relative Resistivity of Various Conductors.

Conductor's material.	Resistivity (Ω.m) at 20°C.
Silver	1.64×10^{-8}
Copper (annealed)	1.724×10^{-8}
Aluminum	2.83×10^{-8}
Tungsten	5.5×10^{-8}
Nickel	7.8×10^{-8}
Iron (pure)	12.0×10^{-8}
Constantan	49.0×10^{-8}

sectional area of 1 square meter. Therefore:

$$\sigma = \frac{1}{\rho}$$

and is measured in siemens per meter (S/m). For example, σ for iron at 20°C is $1/(12.0 \times 10^{-8}) = 8.33 \times 10^7$ S/m.

Example 2-21

An electrical conductor, 2.5 meters long, has a cross-sectional area of 0.75 square millimeter and a resistance of 0.012 ohm. What is the resistance of 30 meters of wire which is made from the same material and has a cross-sectional area of 0.4 square millimeter?

Solution

The resistance, R, is directly proportional to the conductor's length, L, and inversely proportional to the cross-sectional area, A. Therefore

$$\frac{\text{New R}}{\text{Old R}} = \frac{\text{New L/New A.}}{\text{Old L/Old A.}}$$

$$\frac{\text{New R}}{0.012} = \frac{30/0.4}{2.5/0.75} = \frac{30 \times 0.75}{2.5 \times 0.4} = 22.5$$

New Resistance = $22.5 \times 0.012 = 0.27\,\Omega$.

Example 2-22

What is the resistance of a nickel conductor which is 3 meters long and has a cross-sectional area of 2 square millimeters?

Solution

$A = 2 \times 10^{-6}\,m^2$, $L = 3\,m$, $\rho = 7.8 \times 10^{-8}\,\Omega.m$.

$$\text{Resistance} = \rho \frac{L}{A} = 7.8 \times 10^{-8} \,\Omega.\text{m} \times \frac{3m}{2 \times 10^{-6}\text{m}^2} = 0.117\,\Omega.$$

Example 2-23

A 50 meter length of annealed copper wire has a circular cross-sectional area whose diameter is 0.85 mm. What is its resistance at 20°C?

Solution

$$\text{Cross-sectional area, } A = \frac{\pi \, d^2}{4} = \frac{\pi}{4} \times (0.85 \times 10^{-3})^2$$

$$= 5.67 \times 10^{-7}\text{m}^2, \; L = 50\,\text{m}, \; \rho = 1.724 \times 10^{-8}\,\Omega.\text{m}.$$

$$\text{Resistance} = \rho \frac{L}{A} = 1.724 \times 10^{-8}\Omega.\text{m} \times \frac{50\,\text{m}}{5.67 \times 10^{-7}}$$

$$= 1.52\,\Omega.$$

Example 2-24

A tungsten filament has a length of 3 inches and a cross-sectional area of 2 circular mils. What is its resistance at 20°C?

Solution

$L = 3/12 = 0.25$ Ft; $\rho = 5.5 \times 10^{-8} \times 6.015 \times 10^8 = 33.1\Omega$ cmil/ft.

$$\text{Resistance} = \rho \frac{L}{A} = 33.1 \times \frac{0.25}{2} = 4.14\,\Omega.$$

TEMPERATURE COEFFICIENT OF RESISTANCE

For most conducting materials such as copper and aluminum the resistance rises in a linear manner with an increase of temperature over normal temperature ranges. By contrast there are some alloys, for example eureka (60% copper, 40% nickel), whose resistance is little affected by temperature. Finally there are some elements (carbon, germanium, silicon) where there is a reduction in resistance as the temperature increases.

The change in resistance with temperature is measured by the temperature coefficient of resistance, α (alpha). This is commonly defined as the ohmic increase per ohm of resistance at 20°C per degree centigrade (Celsius) rise in temperature. For example, α at 20°C for copper is + 0.00393 $\Omega/\Omega/°C$; this means that if a length of

Table 2-4. Resistance Temperature Coefficients.

Conductor material	$\alpha(20\%C)$
Silver	0.0038
Aluminum	0.0039
Copper	0.00393
Tungsten	0.0045
Iron	0.0055
Nickel	0.006
Constantan	0.0000008
Carbon	−0.0005

copper wire has a resistance of one ohm at 20°C, the resistance will increase by 0.00393 Ω for every 1°C rise in temperature. The values of the temperature coefficient for various materials are shown in Table 2-4; notice that constantan is an alloy with a very low temperature coefficient and that carbon has a negative coefficient. Also, see Table 2-5.

In equation form: $R_{T°} = R_{20°C}(1 + \alpha_{20°C}(T° − 20°))$ or $R_{20°C} = R_{T°}/(1 + \alpha_{20°C}(T° − 20°))$

Table 2-5. Electrical Units.

QUANTITY	UNIT	UNIT SYMBOL	LETTER SYMBOL
CURRENT	AMPERE	A	I
CHARGE	COULOMB, AMPERE-HOUR	C,Ah	Q
EMF or p.d	VOLT	V	E. V
WORK, ENERGY	JOULE, WATT-HOUR, KILOWATT HOUR	J,Wh kWh	W
POWER	WATT	W	P
RESISTANCE	OHM	Ω	R
CONDUCT-ANCE	SIEMENS (MHO)	S	G
SPECIFIC RESISTANCE RESISTIVITY	OHM.METER	Ω.m	ρ
CONDUCT-IVITY	SIEMENS PER METER	S/M	σ
TEMPER-ATURE COEFFIC-IENT OF RESIST-ANCE	OHMS PER OHMS PER DEGREE CENT-GRADE	Ω/Ω°C	α

Where $R_{T°}$ = conductor's resistance at T°C.

$R_{20°C}$ = conductor's resistance at 20°C.

α = temperature coefficient at 20°C.

Since $R_{T°} = \rho_{20°C} \times \dfrac{L}{A}$

$$R_{T°} = \rho_{20°C} \times \dfrac{L}{A} (1 + \alpha_{20°C} (T° - 20°)).$$

and $\rho_{T°} = \rho_{20°} (1 + \alpha_{20°C}(T° - 20°))$.

Example 2-25

A length of copper wire has a resistance of 25 ohms at 20°C. Calculate its resistance at (a) 100°C and (b) 0°C.

Solution

$$\begin{aligned}
R_{100°C} &= 25 (1 + 0.00393 \times (100° - 20°)) \\
&= 25 \times (1 + 0.00393 \times 80) \\
&= 25 \times 1.3144 \\
&= 32.86 \ \Omega. \\
R_{0°C} &= 25 (1 + 0.00393 \times (0° - 20°)) \\
&= 25 (1 - 0.00393 \times 20) \\
&= 25 \times 0.9214 \\
&= 23.04 \ \Omega.
\end{aligned}$$

Example 2-26

Calculate the resistance of 250 meters of nickel wire with a cross-sectional area of 1.75 square millimeters at 60°C.

Solution

$\rho_{20°C} = 7.8 \times 10^{-8} \Omega.m$, $\alpha_{20°C} = 0.006$, $A = 1.75 \times 10^{-6}$ square meter.

Then $R_{60°C} = \dfrac{7.8 \times 10^{-8} \times 250}{1.75 \times 10^{-6}}$ $(1 + 0.006 \times (60° - 20°))$

$$= \dfrac{7.8 \times 10^{-2} \times 250 \times 1.24}{1.75}$$

$$= 13.8 \ \Omega.$$

Example 2-27

What is the resistivity of iron at 80°C?

Solution

$$\rho_{20°C} = 12.0 \times 10^{-8} \, \Omega.m. \; \alpha_{20°C} = 0.0055.$$
$$\rho_{80°C} = 12.0 \times 10^{-8} (1 + 0.0055 \times (80° - 20°))$$
$$= 12.0 \times 10^{-8} \times 1.33$$
$$= 16.0 \times 10^{-8} \, \Omega.m.$$

Example 2-28

A tungsten filament has a resistance of 25Ω at 2000°C. What is its resistance at room temperature (20°C)?

Solution

$$\rho_{20°C} = 0.0045 \, \Omega/\Omega/°C$$
$$R_{20°C} = 25/(1 + 0.0045 \times (2000 - 20))$$
$$= 25/(1 + 0.0045 \times 1980)$$
$$= 25/9.91$$
$$= 2.52 \, \Omega.$$

CHAPTER SUMMARY

☐ Q (coulombs) = I (amperes) × t (seconds).

☐ Mass of substance liberated from electrolyte
$$= z \, (kg/C) \times I \; (amperes) \times t \; (seconds)$$
$$= z \, (kg/C) \times Q \; (coulombs).$$

☐ Electrical Energy or Work = Pt joules.

☐ Current, $I = \dfrac{V}{R} = \dfrac{P}{V} = \sqrt{\dfrac{P}{R}}$ amperes.

☐ Voltage, $V = \dfrac{P}{I} = IR = \dfrac{P}{I} = \sqrt{P \times R}$ volts.

☐ Power, $P = V \times I = I^2 R = \dfrac{E^2}{R}$ watts.

☐ Resistance, $R = \dfrac{V}{I} = \dfrac{P}{I^2} = \dfrac{E^2}{P}$ ohms.

☐ Conductance, $G = \dfrac{1}{R} = \dfrac{I}{V} = \dfrac{I^2}{P} = \dfrac{P}{E^2}$ siemens.

☐ Resistance of cylindrical conductor
$$= \dfrac{P \text{(resistivity)}}{\text{in ohm meter}} \times \dfrac{L \text{ (meters)}}{A \text{ (square meters)}}$$

☐ Resistance of conductor at T°C = $R_{20°} (1 + \alpha_{20°C}(T° - 20°))$.

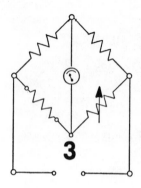

Resistor Networks

Referring to Fig. 3-1, the resistors R_1, R_2 .... R_N are connected in series because they are joined end-to-end in succession. The properties of this circuit are:

    1. There is only a single path for current flow and the current has the same value throughout the circuit.

    2. The sum of the individual voltage drops across the resistors is equal to the value of the source voltage (*Kirchhoff's Voltage Law*).

RESISTORS IN SERIES

    For N resistors in series: $E_T = V_1 + V_2 + V_3 + \ldots + V_N$.

    And: $V_1 = IR_1$, $V_2 = IR_2 \ldots V_N = IR_N$.

    Notice that the highest value resistor develops the greatest voltage drop. If the N resistors all have the same resistance R, $E_T = NV$ and $V_1 = V_2 = \ldots = V_N = V = IR$. where V is the voltage drop across each resistor.

    3. The total equivalent resistance is equal to the sum of the individual resistances.

    For N resistors in series,

$$R_T = R_1 + R_2 + R_3 + \ldots + R_N$$

$$I = \frac{E_T}{R_T} = \frac{E_T}{R_T R_1 + R_2 + R_3 + \ldots + R_N}$$

    If the n resistors all have the same resistance R, $R_T = NR$ and $I = E_T/NR$.

4. The total power delivered from the source is equal to the sum of the individual powers dissipated in the resistors.

For N resistors in series,

$$P_T = P_1 + P_2 + P_3 + \ldots + P_N$$

$$= I^2 R_T = \frac{E^2_T}{R_T} = E_T I.$$

where $P_1 = I^2 R_1 = \frac{V_1^2}{R_1} = IV_1$, etc.

Notice that the highest value resistor dissipates the most power.

If the N resistors all have the same resistance, R:

$$P_T = NP \text{ and } P_1 = P_2 = \ldots = P = I^2 R = \frac{V^2}{R} = IV.$$

VOLTAGE DIVISION RULE

5. The source voltage divides between the resistors in *direct* proportion to the values of the resistors (voltage division rule). Then:

$$V_1 = V_2 \times \frac{R_1}{R_2} = E_T \times \frac{R_1}{R_T}, \text{ etc.}$$

For two resistors in series,

$$V_1 = E_T \times \frac{R_1}{R_1 + R_2}$$

and

$$V_2 = E_T \times \frac{R_2}{R_1 + R_2}.$$

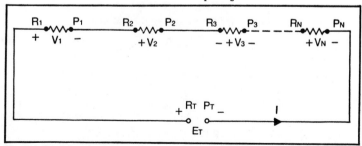

Fig. 3-1. Resistors in series.

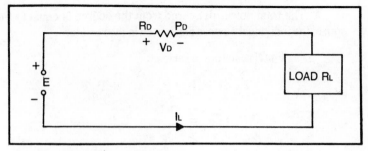

Fig. 3-2. The series dropping resistor.

The total power delivered from the source is also divided between the resistors in direct proportion to the values of the resistors.

$$P_1 = P_2 \times \frac{R_1}{R_2} = P_T \times \frac{R_1}{R_T} \text{, etc.}$$

THE OPEN CIRCUIT

6. An open circuit has theoretically infinite resistance. No current may flow through an open circuit but in a series string of resistors the source voltage appears across the open circuit.

THE SERIES DROPPING RESISTOR

7. A series dropping resistor may be used to reduce a source voltage down to the level required by a load. Referring to Fig. 3-2,

$$V_D = E - V_L = E - I_L \times R_L$$
$$R_D = \frac{V_D}{I_L}$$
$$P_D = I_L \times V_D$$

Example 3-1

In Fig. 3-3 what are the values of the voltages V_{A-B}, V_{D-E}, V_{B-H}, V_{H-P}, V_{N-S}, V_{E-P}, V_{A-S}, V_{Q-J}? (V_{A-B} means the voltage between the points A and B).

Solution

$V_{A-B} = 60\,V$.
$V_{D-E} = 0\,V$. Between D and E there is only a connecting wire across which there is a negligible voltage drop.
$V_{B-H} = V_{B-C} + V_{C-D} + V_{D-E} + V_{E-F} + V_{F-G} + V_{G-H}$

34

$$= 0\,V + 10\,V + 0\,V + 40\,V + 0\,V + 15\,V = \mathit{65\,V}.$$

$$V_{H\text{-}P} = V_{H\text{-}J} + V_{J\text{-}K} + V_{K\text{-}L} + V_{L\text{-}M} + V_{M\text{-}N} + V_{N\text{-}P}$$
$$= 0\,V + 25\,V + 0\,V + 30\,V + 0\,V + 50\,V = \mathit{105\,V}.$$

$$V_{N\text{-}S} = V_{N\text{-}P} + V_{P\text{-}Q} + V_{Q\text{-}S}.$$
$$= 50\,V + 0\,V + 20\,V = \mathit{70\,V}.$$

$$V_{E\text{-}P} = V_{E\text{-}F} + V_{F\text{-}G} + V_{G\text{-}H} + V_{H\text{-}J} + V_{J\text{-}K} + V_{K\text{-}L} + V_{L\text{-}M} + V_{M\text{-}N} + V_{N\text{-}P}$$
$$= 40\,V + 0\,V + 15\,V + 0\,V + 25\,V + 0\,V + 30\,V + 0\,V + 50\,V = \mathit{160\,V}.$$

$V_{A\text{-}S}$ = 250 V. The 250 V battery is connected between the points A and S.

$$V_{Q\text{-}J} = V_{Q\text{-}P} + V_{P\text{-}N} + V_{N\text{-}M} + V_{M\text{-}L} + V_{L\text{-}K} + V_{K\text{-}J}$$
$$= 0\,V + 50\,V + 0\,V + 30\,V + 0\,V + 25\,V = \mathit{105\,V}.$$

Example 3-2

In Fig. 3-4 what are the values of the potential differences $V_{A\text{-}B}$, $V_{B\text{-}E}$, $V_{C\text{-}F}$, $V_{A\text{-}D}$, $V_{D\text{-}F}$?

Solution

Total resistance, R_T = 680 + 220 + 560 + 470 + 270
$$= 2200\,\Omega.$$

Current, $I = \dfrac{E_T}{R_T} = \dfrac{16\,V}{2200\,\Omega} = 0.0072727\,A.$

Potential difference, $V_1 = IR_1 = 0.0072727\,A \times 680\,\Omega$
$$= 4.945436\,V.$$
$$V_2 = IR_2 = 0.0072727\,A \times 220\,\Omega$$
$$= 1.599994\,V.$$
$$V_3 = IR_3 = 0.0072727\,A \times 560\,\Omega$$
$$= 4.072712\,V.$$

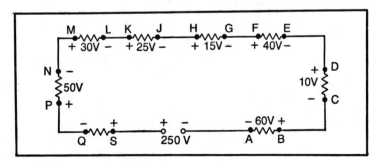

Fig. 3-3. Circuit for example 3-1.

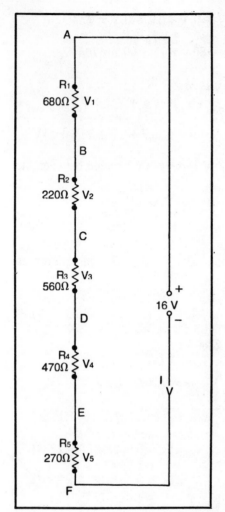

Fig. 3-4. Circuit for example 3-2.

$$V_4 = IR_4 = 0.0072727\,A \times 470\,\Omega$$
$$= 3.418169\,V.$$
$$V_5 = IR_5 = 0.0072727\,A \times 270\,\Omega$$
$$= 1.963629\,V.$$

Voltage Check:

$$
\begin{aligned}
\text{Sum of voltage drops} &= V_1 + V_2 + V_3 + V_4 + V_5 \\
&= 4.945436 + 1.599994 + 4.072712 \\
&\qquad + 3.418169 + 1.963629 \\
&= 15.99994\,V.
\end{aligned}
$$

This compares with the applied voltage of 16 volts; the small difference is due to the rounding off.

$V_{A-B} = V_1 = 4.945436 = $ *4.95 V*, rounded off.
$V_{B-E} = V_2 + V_3 + V_4 = 9.090875 = $ *9.09 V*, rounded off.
$V_{C-F} = V_3 + V_4 + V_5 = 9.45451 = $ *9.45 V*, rounded off.
$V_{C-F} = V_1 + V_2 + V_3 = 10.618142 = $ *10.62 V*, rounded off.
$V_{A-D} = V_4 + V_5 = 5.381798 = $ *5.38 V*, rounded off.

Example 3-3

Find the values of V_1, V_2, V_3, V_4, V_5, P_1, P_2, P_3, P_4, P_5 and the total power, P_T dissipated in the circuit of Fig. 3-5.

Solution

Total resistance, $R_T = 2.2 + 6.8 + 4.7 + 3.3 + 1.0$
$= 18 \text{ k}\Omega.$

The circuit current, $I = \dfrac{E_T}{R_T} = \dfrac{72\text{ V}}{18\text{ K}\Omega} = 4 \text{ mA}.$

The individual voltage drops are
$V_1 = IR_1 = 4\text{ mA} \times 2.2\text{ k}\Omega = $ *8.8 V*.
$V_2 = IR_2 = 4\text{ mA} \times 6.8\text{ k}\Omega = $ *27.2 V*.
$V_3 = IR_3 = 4\text{ mA} \times 4.7\text{ k}\Omega = $ *18.8 V*.
$V_4 = IR_4 = 4\text{ mA} \times 3.3\text{ k}\Omega = $ *13.2 V*.
$V_5 = IR_5 = 4\text{ mA} \times 1.0\text{ k}\Omega = $ *4.0 V*.

The sum of the voltage drops is $V_1 + V_2 + V_3 + V_4 + V_5 = 8.8 + 27.2 + 18.8 + 13.2 + 4.0 = 72.0 \text{ V}$. which exactly balances the source voltage, E_T.

The individual powers dissipated in the resistors are:

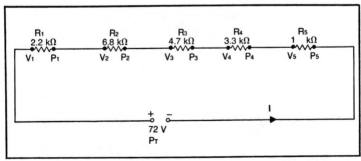

Fig. 3-5. Circuit for example 3-3.

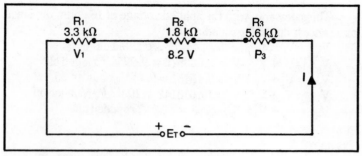

Fig. 3-6. Circuit for example 3-4.

$$P_1 = IV_1 = 4\,mA \times 8.8\,V = 35.2\,mW.$$
$$P_2 = IV_2 = 4\,mA \times 27.2\,V = 108.8\,mW.$$
$$P_3 = IV_3 = 4\,mA \times 18.8\,V = 75.2\,mW.$$
$$P_4 = IV_4 = 4\,mA \times 13.2\,V = 52.8\,mW.$$
$$P_5 = IV_5 = 4\,mA \times 4.0\,V = 16.0\,mW.$$

The total power dissipated by the resistors is

$$P_T = P_1 + P_2 + P_3 + P_4 + P_5$$
$$= 35.2 + 108.8 + 75.2 + 52.8 + 16.0 = 288.0\,mW.$$

The total power delivered by the source is

$E_T \times I = 72\,V \times 4\,mA = 288\,mW$, which exactly balances the value of P_T.

Example 3-4

In Fig. 3-6 calculate the values of V_1, E_T and P_3.

Solution

$$V_1 = V_2 \times \frac{R_1}{R_2} = 8.2 \times \frac{3.3}{1.8} = 15.03\,V.$$

$$E_T = V_t \times \frac{R_T}{R_2} = 8.2 \times \frac{(3.3 + 1.8 + 5.6)}{1.8} = 8.2 \times \frac{10.7}{1.8}$$
$$= 48.74\,V.$$

$$P_3 = P_2 \times \frac{R_3}{R_2} \text{ where } P_2 = \frac{V_2^{\,2}}{R_2} = \frac{(8.2\,V)^2}{1.8\,k\Omega} = 37.36\,mW.$$

$$P_3 = 37.36 \times \frac{5.6}{1.8} = 116.2\,mW$$

38

Example 3-5

Using Fig. 3-7 calculate the values of V_2 and V_3.

Solution

$$V_2 = E_T \times \frac{R_2}{R_T} = 53 \times \frac{1.8}{3.9 + 1.8 + 1.2 + 5.6}$$

$$= 53 \times \frac{1.8}{12.5} = 7.63\,V.$$

$$V_3 = E_T \times \frac{R_3}{R_T} = 53 \times \frac{1.2}{12.5} = 5.09\,V.$$

V_3 could also be calculated from:

$$V_3 = V_2 \times \frac{R_3}{R_2} = 7.63 \times \frac{1.2}{1.8} = 5.09\,V.$$

Example 3-6

In Fig. 3-8 what is the voltage between the points X and Y?

Solution

Since there is an open circuit between X and Y, there is no current flow through the 15 kΩ resistor and therefore no voltage drop across this resistor. Consequently the voltage between X and Y is *10 V.*

Example 3-7

A relay whose coil resistance is 40Ω is designed to function correctly when 2 A flows through the coil. If the relay is operated

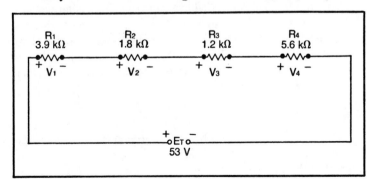

Fig. 3-7. Circuit for example 3-5.

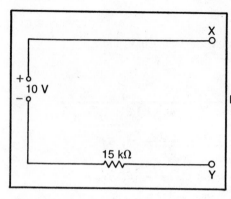

Fig.3-8.Circuit for example 3-6.

from 115 V DC, find the value of the series dropping resistor required and calculate its power dissipation.

Solution

Referring to Fig. 3-2, $R_L = 40\Omega$, $I_L = 2$ A and $E = 115$ V.

$$V_D = E - I_L R_L = 115\,V - 2\,A \times 40\Omega = 115\,V - 80\,V$$
$$= 35\,V.$$

$$R_D = V_D/I_L = 35\,V/2\,A = \mathit{17.5\ \Omega}.$$

$$P_D = I_L \times V_D = 2\,A \times 35\,V = \mathit{70\ W}.$$

GROUND AS A RETURN LINE AND AS A VOLTAGE REFERENCE LEVEL

Ground may be considered as any large mass of conducting material, (for example the metal chassis of a transmitter) so that there is essentially zero resistance between any two ground points. The main reason for using a ground system is to simplify circuitry by saving on the amount of wiring required. Ground is then used as the return path for many circuits so that at any time there are a number of currents flowing through the ground system. However because of the ground's zero resistance property there is no voltage developed between any two ground points and therefore there is no interference between the various circuits. This is illustrated in Figs. 3-9A and 3-9B which show two circuits with four connecting wires. However with a common ground (symbols ⏚ or ⏚) as the return path (Fig. 3-9C), only two connecting wires are needed and there is zero voltage drop between the ground points.

As well as providing a return path, ground is also chosen as a reference level of zero volts and the voltage or potential at any point may then be measured relative to ground. In Fig. 3-10 A, point P is 10 volts positive with respect to point Q and point Q is 10

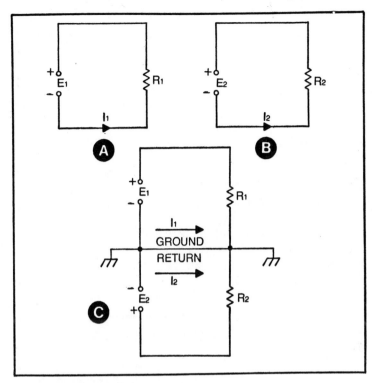

Fig. 3-9. The ground return.

volts negative with respect to point P. However if point Q is grounded at zero volts (Fig. 3-10B), point P carries a voltage of positive 10 volts (+ 10 V), while if P is grounded (Fig. 3-10 C), the potential at Q is negative 10 volts (− 10 V). Individual points may therefore either possess a positive or a negative potential with

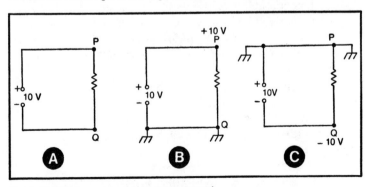

Fig. 3-10. Voltage measured relative to ground.

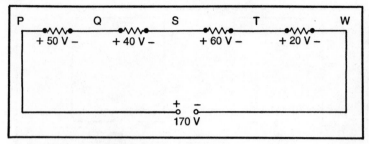

Fig. 3-11. Circuit for example 3-8.

respect to ground. If the electron flow is from ground to a particular point, that point's potential will be positive but if the flow is reversed, the potential is negative.

In any particular circuit the potentials at the various points will depend on which point is grounded; however the voltage drops between any two points will remain the same.

Example 3-8

In the circuit of Fig. 3-11 each of the points is grounded in turn. What voltages with respect to ground then exist at all of the other points in the circuit?

Solution

P grounded.
    Potential at Q = $-$ *50 V.*
    Potential at S = $-$ (50 + 40) = $-$ *90 V.*
    Potential at T = $-$ (50 + 40 + 60) = $-$ *150 V.*
    Potential at W = $-$ (50 + 40 + 60 + 20) = $-$ *170 V.*
Q grounded.
    Potential at P = $+$*50 V.*
    Potential at S = $-$ *40 V.*
    Potential at T = $-$ (40 + 60) = $-$ *100 V.*
    Potential at W = $-$ (40 + 60 + 20) = $-$ *120 V.*
S grounded.
    Potential at P = $+$ (40 + 50) = $+$ *90 V.*
    Potential at Q = $+$*40 V.*
    Potential at T = $-$ *60 V.*
    Potential at W = $-$ (60 + 20) = $-$ *80 V.*
T grounded.
    Potential at P = $+$ (60 + 40 + 50) = *150 V.*
    Potential at Q = $+$ (60 + 40) = $+$ *100 V.*

Potential at S = + *60 V.*
Potential at W = − *20 V.*
W grounded.
Potential at P = + (20 + 60 + 40 + 50) = + *170 V.*
Potential at Q = + (20 + 60 + 40) = +*120 V.*
Potential at S = + (20 + 60) = *80 V.*
Potential at T = + *20 V.*

RESISTORS IN PARALLEL

Referring to Fig. 3-12 A the resistors R_1, R_2 ... R_N connected in parallel because they are joined between two common points X

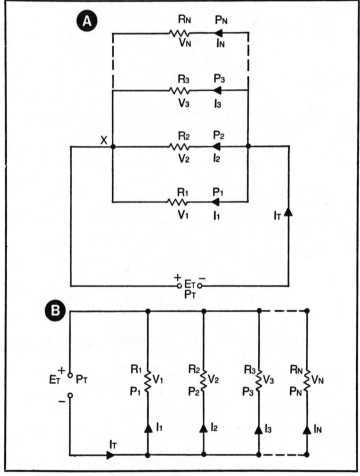

Fig. 3-12. Resistors in parallel.

and Y. Electronically this is the same as connecting the resistors between two common lines as shown in Fig. 3-12 B. The properties of this circuit are:

1. The voltage drop across each resistor is the same and equal to the source voltage.

$$E_T = V_1 = V_2 = \text{----} = V_N.$$

2. The sum of the individual branch currents flowing through the resistors is equal to the total current drawn from the source voltage (*Kirchhoff's Current Law*).

For N resistors is parallel:

$$I_T = I_1 + I_2 + I_3 + \text{----} + I_N$$

where: $I_1 = \dfrac{V_1}{R_1} = \dfrac{E_T}{R_1}$, $I_2 = \dfrac{V_2}{R_2} = \dfrac{E_T}{R_2}$ ----

$$I_N = \dfrac{V_N}{R_N} = \dfrac{E_T}{R_N} .$$

Notice that the lowest value resistor carries the largest current. If the N resistors all have the same resistance, R.

$$I_T = N I \text{ and } I_1 = I_2 = \text{---} = I_N = I = \dfrac{E_T}{R}$$

where I is the branch current through each resistor.

3. The total equivalent conductance is equal to the sum of the individual conductances.

For N resistors in parallel,

$$G_T = G_1 + G_2 + G_3 + \text{---} + G_N$$

$$\dfrac{1}{R_T} = \dfrac{1}{R_1} + \dfrac{1}{R_2} + \dfrac{1}{R_3} + \text{----} + \dfrac{1}{R_N} \text{ (reciprocal formula)}$$

and $I_T = \dfrac{E_T}{R_T} = E_T G_T = E_T \times (G_1 + G_2 + G_3 + ... + G_N)$

Notice that the total equivalent resistance, R_T, is always less than that of the lowest value resistor in parallel.

For two resistors in parallel

$$R_T = \dfrac{R_1 \times R_2}{R_1 + R_2} \text{ (product-over-sum formula)}$$

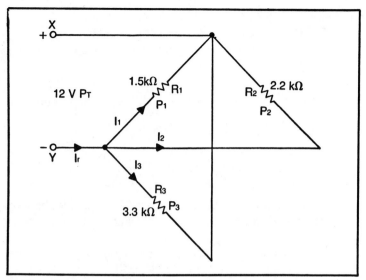

Fig. 3-13. Circuit for example 3-9.

If the N resistors all have the same resistance, $R = 1/G$.

$$G_T = NG, R_T = \frac{R}{N} \text{ and } I_T = N \times \frac{E_T}{R}$$

-4. The total power delivered from the source is equal to the total sum of the powers dissipated in the individual resistors. For N resistors in parallel,

$$P_T = P_1 + P_2 + P_3 + \dots + P_N$$

$$= I_T^2 R_T \frac{E_T^2}{R_T} = E_T I_T$$

$$\text{where } P_1 = \frac{E_T^2}{R_1} = I_1^2 R_1 = E_T I_1, \text{ etc.}$$

Notice that the lowest value resistor dissipates the most power. If the N resistors all have the same resistance R,

$$P_T = NP \text{ and } P_1 = P_2 = \dots = P_N = P = I^2 R = \frac{E_T^2}{R} = E_T I$$

45

CURRENT DIVISION RULE

5. The total current drawn from the source divides between the resistors in *inverse* proportion to the values of the resistors (Current Division Rule).

Then: $I_1 = I_2 \times \dfrac{R_2}{R_1} = I_T \times \dfrac{R_T}{R_1} = I_T \times \dfrac{G_1}{G_T}$, etc

For two resistors in parallel:

$$I_1 = I_T \times \dfrac{R_2}{R_1 + R_2} , I_2 = I_T \times \dfrac{R_1}{R_1 + R_2}$$

The total power delivered from the source is also divided between the resistors in inverse proportion to the values of the resistors.

$$P_1 = P_2 \times \dfrac{R_2}{R_1} = P_T \times \dfrac{R_T}{R_1} , \text{etc}$$

THE SHORT CIRCUIT

6. The short circuit has theoretically zero resistance so that current may flow through the short circuit without developing any voltage drop. However, resistors themselves rarely become shorted and the short circuit normally occurs as the result of bare connecting wires coming into contact or being joined together by a stray drop of solder. The short circuit is therefore a zero resistance path across, or in parallel with, the resistor.

Example 3-9

In Fig. 3-13 what is the value of the resistance between the points X and Y? What are the values of I_1, I_2, I_3, I_T, P_1, P_2, P_3 and P_T?

Solution

All three resistors are effectively connected between X, Y and are therefore in parallel. To use the product-over-sum method for calculating the equivalent resistance, R_T, of three resistors in parallel, first obtain the equivalent resistance, R^1, of R_1 and R_2 in parallel. Then calculate R_T by finding the total resistance of R^1 and R_3 in parallel.

$$R^1 = \dfrac{R_1 \times R_2}{R_1 + R_2} = \dfrac{1.5 \times 2.2}{1.5 + 2.2} = \dfrac{1.5 \times 2.2}{3.7} = 0.892 \, k\Omega$$

Total resistance between points X and Y,

$$R_T = \frac{R^1 \times R_3}{R^1 + R_3} = \frac{0.892 \times 3.3}{3.3 + 0.892} = \frac{0.892 \times 3.3}{4.192} = 0.702\,k\Omega = 702\,\Omega$$

Alternatively, the reciprocal formula may be used.

$$\frac{1}{R_T} = \frac{1}{1.5} + \frac{1}{2.2} + \frac{1}{3.3}$$

$$= 0.667 + 0.455 + 0.303 = 1.425$$

$$R_T = \frac{1}{1.425} = 0.702\,k\Omega = 702\,\Omega$$

The individual branch currents are

$$I_1 = \frac{E_T}{R_1} = \frac{12\,V}{1.5\,k\Omega} = 8.0\,mA$$

$$I_2 = \frac{E_T}{R_2} = \frac{12\,V}{2.2\,k\Omega} = 5.45\,mA$$

$$I_3 = \frac{E_T}{R_3} = \frac{12\,V}{3.3\,k\Omega} = 3.64\,mA$$

$$I_T = I_1 + I_2 + I_3 = 8.0\,mA + 5.45\,mA + 3.64\,mA = 17.09\,mA$$

Total resistance check. The total resistance is also given by,

$$R_T = \frac{E_T}{I_T} = \frac{12\,V}{17.09\,mA} = 0.702\,k\Omega = 702\,\Omega$$

The total resistance may therefore be found either by repeated use of the product-over-sum formula or by the reciprocal method or by obtaining the total current from the branch currents (if a supply voltage is not given, any convenient value may be assumed).

$$P_1 = E_T I_1 = 12\,V \times 8.0\,mA = 96\,mW$$
$$P_2 = E_T I_2 = 12\,V \times 5.45\,mA = 65.4\,mW$$
$$P_3 = E_T I_3 = 12\,V \times 3.64\,mA = 43.7\,mW$$

Total Power, $P_T = P_1 + P_2 + P_3 = 96\,mW + 65.4\,mW + 43.7\,mW = 205.1\,mW$

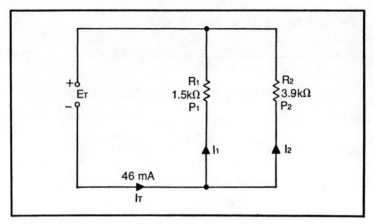

Fig. 3-14. Circuit for example 3-10.

Check: Total power delivered from source,
$$P_T = EI_T = 12\,V \times 17.09\,mA = 205.1\,mW$$

Example 3-10

In the circuit of Fig. 3-14, what are the values of I_1, P_1, P_2 and E_T?

Solution

$$I_1 = I_T \times \frac{R_2}{R_1 + R_2} = 46\,mA \times \frac{3.9\,k\Omega}{1.5\,k\Omega + 3.9\,k\Omega}$$

$$= 46 \times \frac{3.9}{5.4} = 33.2\,mA$$

$$P_1 = I_1^2 R_1 = (33.2\,mA)^2 \times 1.5\,k\Omega = 1656\,mW$$

$$P_2 = P_1 \times \frac{R_1}{R_2} = 1656\,mW \times \frac{1.5\,k\Omega}{3.9\,k\Omega} = 637\,mW$$

Note that the lower value resistor, R_1, dissipates the greater power.

$$E_T = I_1 R_1 = 32.2\,mA \times 1.5\,k\Omega = 50\,V, \text{ rounded off.}$$

Example 3-11

A purely resistive load has a value of 3.7 kΩ which is reduced to 2.5 kΩ by connecting a single resistor across the load. What is the value of this resistor?

48

Solution

The resistor, R_X, is connected in parallel with the load, R_L, so that their combination has a total resistance of $2.5\,k\Omega$
This leads to:

$$\frac{1}{R_T} = \frac{1}{R_L} + \frac{1}{R_X}$$

This leads to:

$$R_X = \frac{R_L R_T}{R_L - R_T} = \frac{2.5 \times 3.7}{3.7 - 2.5}$$

$$= \frac{2.5 \times 3.7}{1.2} = 7.7\,k\Omega$$

Example 3-12

Eight electric light bulbs, each with a rating of 5 W, 110 V, are being operated from a 110 V DC source. What is the value of the total load resistance?

Solution

The "hot" resistance of each bulb $= \dfrac{E^2}{P} = \dfrac{110^2}{75}\ \Omega$ Since the bulbs must be connected in parallel, the total load resistance is

$$\frac{110^2/75}{8} = 20\,\Omega, \text{ rounded off.}$$

Example 3-13

A 20,000 ohm 200 watt resistor, a 40,000 ohm 100 watt resistor and a 5,000 ohm 50 watt resistor are connected in parallel. What is the maximum value of the total source current which will not cause the wattage rating of any of the resistors to be exceeded?

Solution

Since $E = \sqrt{PR}$, the maximum voltage ratings for the individual resistors are $\sqrt{20,000 \times 200}$, $\sqrt{40,000 \times 100}$ and $\sqrt{5,000 \times 50}$. The smallest of these ratings is $\sqrt{5,000 \times 50} = \sqrt{250,000} = \sqrt{25 \times 10^4} = 10^2 \times \sqrt{25} = 500\,V$.

$$\text{Total current} = \frac{500}{20,000} + \frac{500}{40,000} + \frac{500}{5,000}$$

$$= 0.025 + 0.0125 + 0.1$$

$$= 0.1375\,A = 137.5\,mA$$

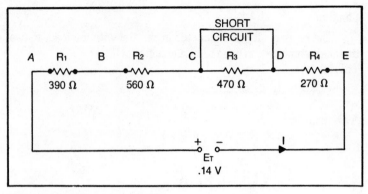

Fig. 3-15. Circuit for example 3-14.

Example 3-14

In Fig. 3-15 point E is grounded and a short appears across the 470Ω resistor. What are the new potentials at the points A, B, C and D?

Solution

Due to the short circuit the resistance between the points C and D is zero. The new total equivalent resistance is:

$$R_T = 390 + 560 + 270 = 1220\Omega.$$

$$\text{Current } I = \frac{E_T}{R_T} = \frac{14\,V}{1220\Omega} = 0.0114754\,A.$$

$$V_1 = IR_1 = 0.0114754\,A \times 390\Omega = 4.475\,V.$$
$$V_2 = IR_2 = 0.0114754\,A \times 560\Omega = 6.426\,V.$$
$$V_3 = 0\,V.$$
$$V_4 = IR_4 = 0.0114754\,A \times 270\Omega = 3.099\,V.$$

Potential at point A = $+14\,V$.

Potential at point B = $+14 - 4.475 = +9.525\,V$.

Potential at points C and D = $+9.525 - 6.426 = +3.099\,V$.

THE VOLTAGE DIVIDER CIRCUIT

This circuit is a practical result of the voltage division rule. By its use a number of different voltages may be made available from a single voltage source.

In Fig. 3-16, V_1 is the voltage drop across the series combination of R_1, R_2, R_3, R_4, R_5 and is therefore equal to the source voltage, E_T. V_2 is dropped across R_5, R_3, R_4, R_5 in series;

$$V_2 = E_T \times \frac{(R_2 + R_3 + R_4 + R_5)}{R_T} = E_T \times \frac{(R_2 + R_3 + R_4 + R_5)}{(R_1 + R_2 + R_3 + R_4 + R_5)}$$

Similarly:

$$V_3 = E_T \times \frac{(R_3 + R_4 + R_5)}{R_T} = E_T \times \frac{(R_3 + R_4 + R_5)}{(R_1 + R_2 + R_3 + R_4 + R_5)}$$

$$V_4 = E_T \times \frac{(R_4 + R_5)}{R_T} = E_T \times \frac{(R_4 + R_5)}{(R_1 + R_2 + R_3 + R_4 + R_5)}$$

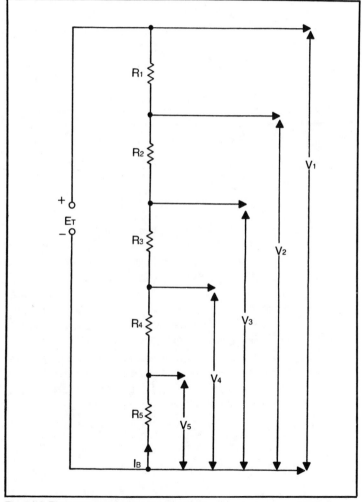

Fig. 3-16. The unloaded voltage divider circuit.

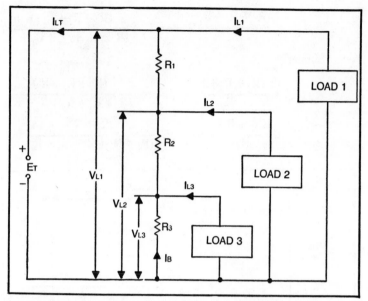

Fig. 3-17. The loaded voltage divider circuit.

and

$$V_5 = E_T \times \frac{R_5}{R_T} = E_T \times \frac{R_5}{(R_1 + R_2 + R_3 + R_4 + R_5)}$$

The voltages V_1, V_2, V_3, V_4, V_5 are developed by the flow of the bleeder current, I_B, through the resistors. The availability of the different output voltages is therefore achieved at the expense of the power dissipated in the series string.

The voltage divider of Fig. 3-16 is operating under no-load conditions. As soon as loads are connected across the voltages V_1, V_2, V_3, V_4, V_5, the circuit becomes a series-parallel arrangement and the four preceding equations are no longer valid.

Figure 3-17 illustrates a loaded voltage divider circuit. The three load voltages are V_{L1}, V_{L2}, V_{L3} and these are respectively associated with the load currents, I_{L1}, I_{L2}, I_{L3}. The bleeder current, I_B, is typically 10% of the total load current, $I_{LT} = I_{L1} + I_{L2} + I_{L3}$. Then

$$V_{L3} = I_B R_3 \text{ and } R_3 = \frac{V_{L3}}{I_B}$$

$$V_{L2} = V_{L3} + R_2(I_{L3} + I_B)$$

$$\text{and } R_2 = \frac{V_{L2} - V_{L3}}{I_{L3} + I_B}$$

$$E_T = V_{L1} = V_{L2} + R_1 (I_{L2} + I_{L3} + I_B)$$

$$R_1 = \frac{V_{L1} - V_{L2}}{I_{L2} + I_{L3} + I_B}.$$

Notice that I_{L1} is drawn directly from the source voltage and therefore does not appear in the equations.

Given the values of the load voltages with their currents and assuming a certain percentage for the bleeder current, the values of R_1, R_2, R_3 may be calculated.

Example 3-15

In Fig. 3-18 find the values of V_1, V_2, V_3, V_4 and V_5.

Solution

$V_1 = 75V.$

$V_2 = 75 \times \dfrac{(100 + 20 + 4 + 1)}{(500 + 100 + 20 + 4 + 1)} = 75 \times \dfrac{125}{625} = 15V.$

$V_3 = 75 \times \dfrac{(20 + 4 + 1)}{(500 + 100 + 20 + 4 + 1)} = 75 \times \dfrac{25}{625} = 3V.$

$V_4 = 75 \times \dfrac{(4 + 1)}{(500 + 100 + 20 + 4 + 1)} = 75 \times \dfrac{5}{625} = 0.6V.$

$V_5 = 75 \times \dfrac{1}{(500 + 100 + 20 + 4 + 1)} = 75 \times \dfrac{1}{625} = 0.12V.$

Fig. 3-18 could be referred to as a "divide-by-5" circuit since V_2 is 1/5th of V_1, V_3 is 1/5th of V_2 etc.

Example 3-16

Referring to Fig. 3-17, $V_{L1} = E_T = 24\,V$, $V_{L2} = 15\,V$, $V_{L3} = 9\,V$, $I_{L3} = 50\,mA$, $I_{L2} = 85\,mA$ and $I_{L1} = 125\,mA$. If the bleeder current is 10 % of the total load current, what are the values required for R_1, R_2 and R_3?

Solution

Total load current, $I_{LT} = I_{L1} + I_{L2} + I_{L3} = 125 + 85 + 50$
$= 260\,mA.$

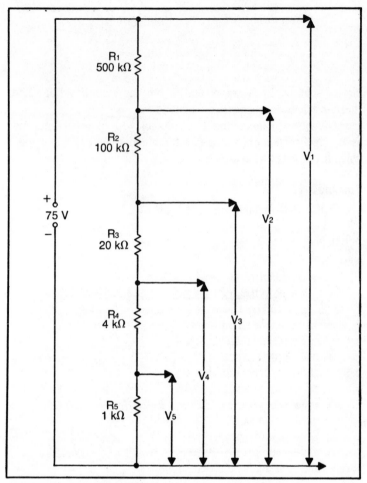

Fig. 3-18. Circuit for example 3-15.

Bleeder current, $I_B = \dfrac{10}{100} \times 260 = 26\,\text{mA}$.

$$R_3 = \frac{V_{L3}}{I_B} = \frac{9\,\text{V}}{26\,\text{mA}} = 346\,\Omega$$

$$R_2 = \frac{V_{L2} - V_{L3}}{I_{L3} + I_B} = \frac{15\,\text{V} - 9\,\text{V}}{50\,\text{mA} + 26\,\text{mA}} = \frac{6\,\text{V}}{76\,\text{mA}} = 79\,\Omega$$

$$R_1 = \frac{V_{L1} - V_{L2}}{I_{L2} + I_{L3} + I_B} = \frac{24\,\text{V} - 15\,\text{V}}{85\,\text{mA} + 50\,\text{mA} + 26\,\text{mA}} = \frac{9\,\text{V}}{161\,\text{mA}} = 56\,\Omega.$$

The power dissipated in these resistors is:

$$P_1 = 9\,V \times 161\,mA = 1449\,mW$$
$$P_2 = 6\,V \times 76\,mA = 456\,mW$$
$$P_3 = 9\,V \times 26\,mA = 234\,mW$$

Suitable resistors would be R_1 56 Ω 2 W, R_2 82 Ω 1 W and R_3 330 Ω ½ W.

SERIES-PARALLEL RESISTOR NETWORKS

A series-parallel circuit may be defined as one in which parts of the circuit have the properties of a simple series arrangement while other parts have the properties of a basic parallel connection. There are an infinite variety of such circuits but one example is shown in Fig. 3-19 A. Its analysis may be carried out in a number of steps.

Step 1. Identify all single resistors which are directly in series with each other, or in other words, those resistors which carry the same current. Combine all such series resistors into their equivalent resistances. In the example chosen, R_2 and R_3 are in series and their total resistance is 11 kΩ + 4 kΩ = 15 kΩ. As far as the equivalent resistance is concerned, the order in which the components are connected is unimportant; therefore R_4, R_5 are also in series with a total resistance of 1 kΩ + 3 kΩ + = 4 kΩ. The circuit may now be redrawn as shown in Fig. 3-19 B.

Step 2. Identify all single resistors which are directly in parallel with other. These resistors will be connected between points where a current division occurs, so that R_6, R_7 and R_8 are in parallel and their equivalent resistance, R_{eq}, is given by:

$$\frac{1}{R_{eq}} = \frac{1}{6} + \frac{1}{3} + \frac{1}{2} = \frac{1}{1}$$
$$R_{eq} = 1\,k\Omega.$$

The circuit may now be drawn again as in Fig. 3-19 C.

Step 3. Combine all equivalent resistances which are in series. This applies to the equivalent resistances of R_4, R_5 and R_6, R_7, so that the total resistance of these five resistors is 4 kΩ + 1k Ω = 5kΩ. Again the circuit is redrawn as in Fig. 3-19D.

Step 4. Combine all equivalent resistances which are in parallel. In Fig. 3-19 D a current split occurs at point X so that the equivalent resistances of R_2, R_3 and R_4, R_5, R_6, R_7, R_8 are in parallel. The total resistance of these seven resistors is

55

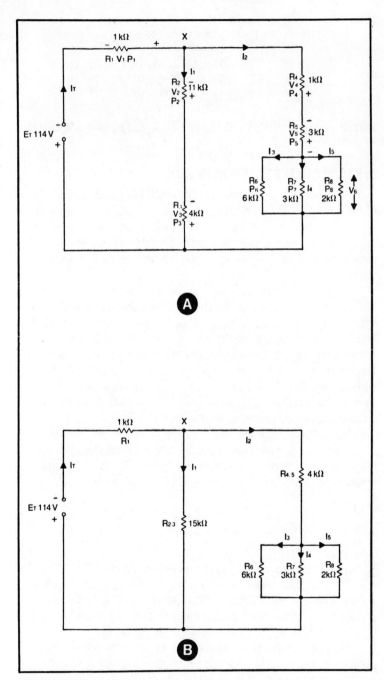

Fig. 3-19 A & B. Combining the series resistors.

$$\frac{5 \times 15}{5 + 15} = \frac{5 \times 15}{20} = 3.75 \, \text{k}\Omega.$$

The final redrawing of the circuit is then shown in Fig. 3-19 E.

Step 5. The total resistance, R_T, of the whole circuit is the series combination of R_1 and the equivalent resistance of R_2, R_3, R_4, R_5, R_6, R_7, R_8. R_T is therefore equal to $1 \, \text{k}\Omega + 3.75 \, \text{k}\Omega = 4.75 \, \text{k}\Omega$.

In carrying out the steps to find R_T, the equivalent resistances of series and parallel combinations were alternately calculated. The circuit was broken down a bit at a time with each redrawing simpler than before. After some experience with series-parallel circuits, the intermediate drawings may be omitted. The tendency with these problems is to start furthest away from the source and gradually work towards the source.

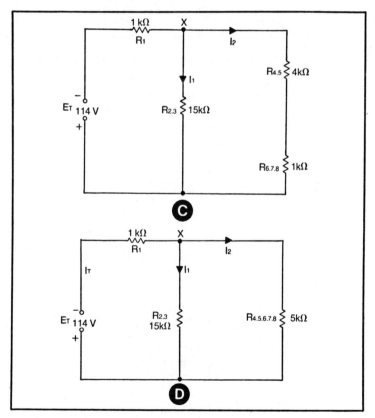

Fig. 3-19 C & D. Combining the parallel resistors.

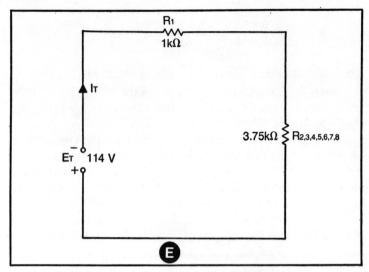

Fig. 3-19 E. Simplified circuit showing the equivalent resistances.

The next stage in the analysis is to work away from the source and determine the individual currents, voltage drops and powers dissipated in the various resistors. When calculating the required values it is important to check that the voltages around any closed loop exactly balance (Kirchhoff's Voltage Law). At the same time the currents existing at any point must also balance (*Kirchhoff's Current Law*).

$$\text{Total current, } I_T = \frac{E_T}{R_T} = \frac{114\,\text{V}}{3.75\,\text{k}\Omega} = 24\,\text{mA}.$$

$$V_1 = I_T R_1 = 24\,\text{mA} \times 1\,\text{k}\Omega = 24\,\text{V}.$$

By Kirchhoff's Voltage Law,

$$V_2 + V_3 = E_T - V_1 = 114\,\text{V} - 24\,\text{V} = 90\,\text{V}.$$

Then:

$$I_1 = \frac{V_2 + V_3}{R_2 + R_3} = \frac{90\,\text{V}}{11\,\text{k}\Omega + 4\,\text{k}\Omega} = 6\,\text{mA}.$$

$$V_2 = I_1 R_2 = 6\,\text{mA} \times 11\,\text{k}\Omega = 66\,\text{V}.$$
$$V_3 = I_1 R_3 = 6\,\text{mA} \times 4\,\text{k}\Omega = 24\,\text{V}.$$

By Kirchhoff's Current Law:

$$I_2 = I_T - I_1 = 24\,\text{mA} - 6\,\text{mA} = 18\,\text{mA}.$$

Then:

$$V_4 = I_2 R_4 = 18\,\text{mA} \times 1\,\text{k}\Omega = 18\,\text{V}.$$
$$V_5 = I_2 R_5 = 18\,\text{mA} \times 3\,\text{k}\Omega = 54\,\text{V}.$$

By Kirchhoff's Voltage Law:

$$V_6 = 90\,\text{V} - V_4 - V_5 = 90\,\text{V} - 18\,\text{V} - 54\,\text{V} = 18\,\text{V}.$$

Then:

$$I_3 = \frac{V_6}{R_6} = \frac{18\,\text{V}}{6\,\text{k}\Omega} = 3\,\text{mA}.$$

$$I_4 = \frac{V_6}{R_7} = \frac{18\,\text{V}}{3\,\text{k}\Omega} = 6\,\text{mA}.$$

$$I_5 = \frac{V_6}{R_8} = \frac{18\,\text{V}}{2\,\text{k}\Omega} = 9\,\text{mA}.$$

$$P_1 = I_T V_1 = 24\,\text{mA} \times 24\,\text{V} = 576\,\text{mW}.$$
$$P_2 = I_1 V_2 = 6\,\text{mA} \times 66\,\text{V} = 396\,\text{mW}.$$
$$P_3 = I_1 V_3 = 6\,\text{mA} \times 24\,\text{V} = 144\,\text{mW}.$$
$$P_4 = I_2 V_4 = 18\,\text{mA} \times 18\,\text{V} = 324\,\text{mW}.$$
$$P_5 = I_2 V_5 = 18\,\text{mA} \times 54\,\text{V} = 972\,\text{mW}.$$
$$P_6 = I_3 V_6 = 3\,\text{mA} \times 18\,\text{V} = 54\,\text{mW}.$$
$$P_7 = I_4 V_6 = 6\,\text{mA} \times 18\,\text{V} = 108\,\text{mW}.$$
$$P_8 = I_5 V_6 = 9\,\text{mA} \times 18\,\text{V} = 162\,\text{mW}.$$

Total power dissipated: $P_T = 2736\,\text{mW}$.

Check: Total power delivered from the source is

$$P_T = I_T E_T = 24\,\text{mA} \times 114\,\text{V} = 2736\,\text{mW}.$$

There are many complex resistor networks which may not be treated as simple series-parallel combinations. One example is shown in Fig. 3-20 in which the position of R_5 in relation to the other resistors cannot be specified as either series or parallel. To analyze such a circuit requires some of the more sophisticated techniques which are outlined in Chapter 8.

CHAPTER SUMMARY
□ *Resistors in Series*

$$E_T = V_1 + V_2 + V_3 + \cdots + V_N \ (\text{Kirchhoff's Voltage Law})$$

$$R_T = R_1 + R_2 + R_3 + \cdots + R_N$$

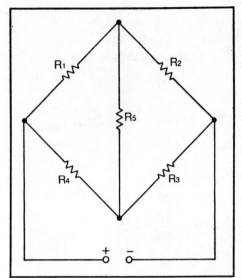

Fig. 3-20. Complex resistor network.

$$P_T = P_1 + P_2 + P_3 + \text{----} + P_N$$

$$V_1 = V_2 \times \frac{R_1}{R_2} = E_T \times \frac{R_1}{R_T}$$

$$P_1 = P_2 \times \frac{R_1}{R_2} = P_T \times \frac{R_1}{R_T}$$

☐ *Series Dropping Resistor*

$$V_D = E - V_L = E - I_L R_L$$

$$R_D = \frac{V_D}{I_L}$$

$$P_D = I_L \times V_D$$

☐ *Resistors in Parallel*

$$I_T = I_1 + I_2 + I_3 + \text{---} + I_N \text{ (Kirchhoff's Current Law)}$$

$$\frac{1}{R_T} = \frac{1}{R_1} + \frac{1}{R_2} + \frac{1}{R_3} + \text{----} + \frac{1}{R_N}$$

$$P_T = P_1 + P_2 + P_3 + \text{----} + P_N$$

$$I_1 = I_2 \times \frac{R_2}{R_1} = \frac{I_T \times R_T}{R_1}$$

$$P_1 = P_2 \times \frac{R_2}{R_1} = P_T \times \frac{R_T}{R_1}$$

For two resistors in parallel:

$$R_T = \frac{R_1 R_2}{R_1 + R_2}$$

$$I_1 = I_T \times \frac{R_2}{R_1 + R_2}, \quad I_2 = I_T \times \frac{R_1}{R_1 + R_2}$$

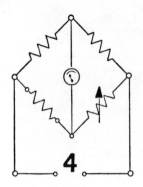

Voltage and Current Sources

Most practical electrical sources such as a battery generate a terminal PD which falls as the load current is increased. This effect can be explained by assuming that the source contains a constant-voltage EMF, E, which is independent of the load current, in series with an internal resistance, R_i. This equivalent circuit is shown in Fig. 4-1. If the load resistance, R_L, is reduced, the load current I_L, will increase. This will raise the voltage drop, V_i, across the internal resistance so that the terminal PD or load voltage, V_L, falls.

CONSTANT VOLTAGE SOURCES

The value of E is found by removing the load and then connecting a voltmeter across the terminals. The voltmeter is assumed to draw negligible current from the source so that the voltage drop across the internal resistance is zero and the terminal PD is equal to the constant voltage EMF. The value of E is therefore referred to as the open-circuit terminal voltage.

If a conductor of negligible resistance is connected between the terminals, R_L is zero and the current will only be limited by the internal resistance. Since the terminals have effectively been short-circuited, the large current that flows is called the short-circuit current whose value is $\frac{E}{R_L}$. The internal resistance is therefore given by:

$$\text{Internal resistance, } R_i = \frac{\text{Open-circuit terminal Voltage, E}}{\text{Short-circuit terminal current, E.}}$$

Normally the short-circuit current is so large that the voltage source will be damaged, if not destroyed. A more practical method

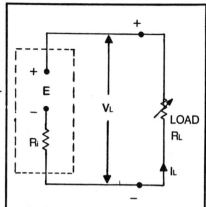

Fig. 4-1. Constant voltage source.

of finding the internal resistance is to adjust the value of R_i until V_L = E/2; R_L and R_i will then be equal.

The equations related to Fig. 4-1 are:

$$\text{Load current, } I_L = \frac{V_L}{R_L} = \frac{V_i}{R_i} = \frac{E}{R_L + R_i}$$

$$\text{Load Voltage, } V_L = I_L R_L = \frac{E \times R_L}{R_L + R_i} = E - V_i = E - I_L R_i$$

Internal voltage drop $= I_L R_i = E - V_L$.

As an example, consider an electrical source with an open circuit terminal voltage of 12V and an internal resistance of $0.15\,\Omega$. Using the equation V_L is calculated for various load currents (Table 4-1).

The graph of load voltage/load current is shown in Fig. 4-2. The graph is a straight line because the internal resistance is constant and may be calculated from the line's negative slope

$$(R_i = \frac{\Delta V_L}{\Delta I_L} = \frac{1.5\,V}{10\,A} = 0.5\,\Omega).$$

In many cases the internal resistance of a power source is not constant and then the V_L/I_L graph will be a regulation curve. The term "voltage regulation" is a measure of the extent to which V_L changes as I_{ol} varies. Voltage regulation is expressed as a percentage and is calculated from:

$$\text{Percentage regulation} = \frac{V_{NL} - V_{FL}}{V_{FL}} \times 100\,\%$$

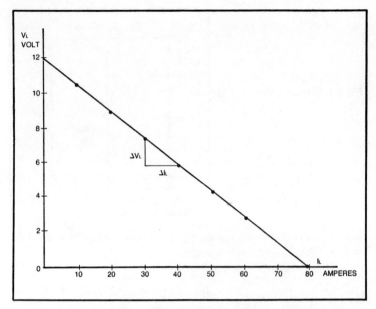

Fig. 4-2. Graph of load voltage/load current from values shown in Table 4-1.

where V_{NL} = Terminal voltage under no-load conditions.

V_{FL} = Terminal voltage under full-load conditions.

Assuming that the open circuit voltage is the no-load condition, $V_{NL} = E$ and $V_{FL} = ER_{FL} / (R_{FL} + R_i)$. Then: Percentage

$$regulation = \frac{\left(E - \frac{ER_{FL}}{R_{FL} + R_i}\right)}{\frac{ER_{FL}}{R_{FL} + R_i}} \times 100 = \frac{R_i}{R_{FL}} \times 100\%$$

where R_{FL} = load resistance corresponding to full-load conditions. In Table 4-1, V_{NL} is 12 V. If the full load current is 20 A, V_{FL} is 9 V.

Then: Percentage regulation $= \frac{12\,V - 9\,V}{9\,V} \times 100 = 33\,\frac{1}{3}\%$.

An ideal voltage source would have zero percentage regulation in which case the regulation "curve" would be a horizontal line. This would indicate that the load voltage was constant and independent to the load current. Therefore, as seen from Table 4-1 the condition for a high load voltage requires that the load resistance is large compared with the internal resistance; typically R_L must be at least five to ten times the value of R_i. The constant voltage EMF will then divide between the internal and load resistances and most of the voltage will appear across the load; by the voltage division rule, $V_L/V_i = R_L/R_i$ and $V_L/E = R_L/(R_L + R_i)$.

Example 4-1

In Fig. 4-1 $E = 42$ V and $R_i = 0.1 \, \Omega$. Under no-load conditions R_L is an open circuit while the full-load value of R_L is $1.5 \, \Omega$. Calculate the full-load values of I_L, V_L, and the percentage of regulation.

Solution

Full-load voltage, $V_{FL} = 42 \, V \times \dfrac{1.5 \, \Omega}{1.5 \, \Omega + 0.1 \, \Omega} = 39.375 \, V.$

Full-load current, $I_L = \dfrac{V_{FL}}{R_L} = \dfrac{39.375 \, V}{1.5 \, \Omega} = 26.25 \, A.$

No-load voltage, $V_{NL} = 42.0 \, V.$

Percentage of regulation $= \dfrac{42.0 - 39.375}{39.375} \times 100 = 6.67\%.$

Check:

Percentage regulation $= \dfrac{R_i}{R_{FL}} \times 100 = \dfrac{0.1}{1.5} \times 100 = 6.67\%.$

CONDITION FOR MAXIMUM POWER TRANSFER

The power in the load, P_L, will be equal to the product of V_L and I_L and is therefore zero under both the open-circuit and the

Table 4-1. Load Voltages and Load Currents.

I_L AMPERES	V_i VOLTS	V_L VOLTS	R_L OHMS
0	0	12	Open circuit. Infinite ohms
10	1.5	10.5	1.05
20	3	9	0.45
30	4.5	7.5	0.25
40	6	6	0.15
50	7.5	4.5	0.09
60	9	3	0.05
80	12	0	Short circuit. Zero ohms

short-circuit conditions. Between these extremes the load power reaches a maximum value corresponding to a particular value of R_L. There are many examples in electronics where it is necessary to know the condition for maximum power transfer to the load. From the following equation: Power developed in the load:

$$P_L = I_L{}^2 R_L = \frac{E}{R_L + R_i}{}^2 \times R_L = \left(\frac{E}{(R_L + R_i)} \right)^2$$

At this point it is customary to use calculus to derive the condition for P_L to reach its maximum value as R_L and hence I_L are varied. However it is possible to obtain the condition by algebra alone. If P_L is to reach its maximum value, its reciprocal $1/P_L$ must at the same time fall to its minimum value. Then:

$$\frac{1}{P_L} = \frac{(R_L + R_i)^2}{E^2 R_L} = \frac{1}{E^2} \left[\frac{(R_L - R_i)^2}{R_L} + 4R_i \right]$$

The minimum value of a square such as $(R_L - R_i)^2$ is zero; this occurs when:

$$R_L = R_i.$$

The condition for maximum power transfer therefore requires that the load resistance, R_L, is *matched* (made equal) to the internal resistance, R_i. As an example a receiver's loadspeaker is matched to the transistor used in the final stage; this enables the audio power developed in this stage to be transferred to the load of the loudspeaker.

When R_L and R_i are matched, the maximum power developed in the load is

$$P_{L\,max} = \frac{E^2}{4 R_i}$$

Another factor which must be taken into account is the circuit efficiency. This is defined as the percentage of the power developed in the load as compared to the total power drawn from the constant voltage EMF.

Therefore:

$$\text{Circuit efficiency} = \frac{I_L{}^2 R_L}{I_L{}^2 (R_L + R_i)} \times 100\% = \frac{R_L}{R_L + R_i} \times 100\%.$$

Since $V_L/E = R_L/(R_L + R_i)$, the condition for a large voltage across the load is the same as for a high circuit efficiency.

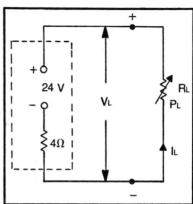

Fig. 4-3. Circuit related to the values shown in Table 4-2.

These results are best illustrated by the example which is shown in Fig. 4-3. The constant voltage EMF is 24 V and this is associated with an internal resistance of 4 Ω. The load resistance, R_L, is varied from zero (short circuit) to infinity (open circuit) and for certain values of R_L, the values of I_L, V_L, P_L and percentage efficiency are calculated as shown in Table 4-2. Figure 4-4 contains the graphs of V_L, I_L, P_L and percentage of efficiency plotted against R_L. Another presentation is in Fig. 4-5, which shows the relationships between I_L and V_L, P_L and percentage of efficiency. The V_L/I_L and percentage of efficiency/I_L graphs are then straight lines while the P_L/I_L curve is a parabola.

Example 4-2

In Fig. 4-1, $E = 36$ V, $R_i = 2.5$ Ω and $R_L = 8$Ω. What additional

Table 4-2. How Load Resistance Affects Efficiency.

R_L OHMS	I_L AMPERES	V_L VOLTS	P_L WATTS	% EFFICIENCY
Open circuit. Infinite Ohms	0	24	0	100
20	1	20	20	83.3
12	1.5	18	27	75
8	2	16	32	66.7
4	3	12	36	50
2	4	8	32	33.3
Short circuit. Zero Ohms.	6	0	0	0

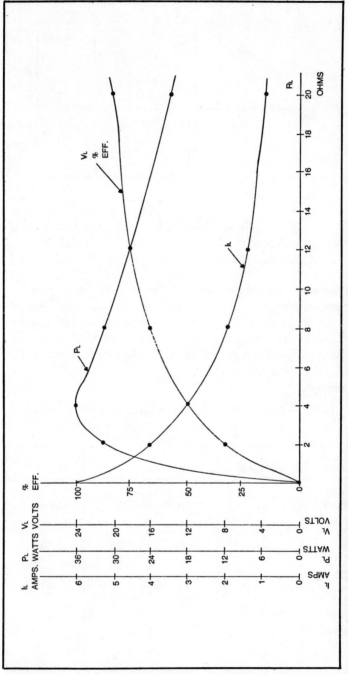

Fig. 4-4. Graphs of load voltage, load current, load power and percentage of efficiency against load resistance.

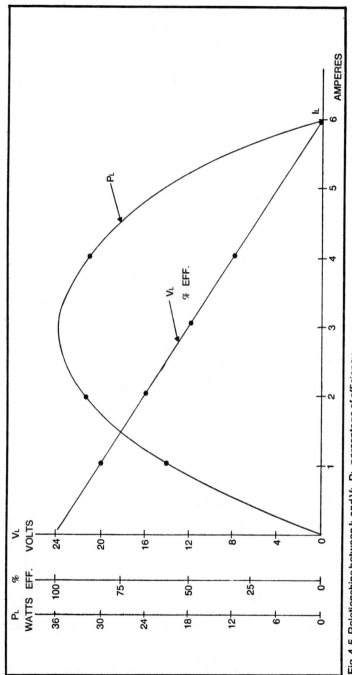

Fig. 4-5. Relationships between I_L and V_L, P_L, percentage of efficiency.

69

load must be added in parallel with R_L in order for maximum power to be developed in the new total load? Calculate the values of P_L before and after the introduction of the additional load.

Solution

For maximum power transfer, the total load must be matched to the internal resistance of $2.5\,\Omega$.

$$\text{Value of the additional load} = \frac{2.5 \times 8}{8 - 2.5} = \frac{20}{5.5} = 3.64\ \Omega.$$

$$\text{Original load power} = \frac{E^2\,R_L}{(R_i + R_L)^2} = \frac{(36\,V)^2 \times 8\,\Omega}{(2.5\,\Omega + 8\,\Omega)^2} = 94.0\ W.$$

$$\text{Final (maximum) load power} = \frac{E^2}{4\,R_i} = \frac{(36\,V)^2}{4 \times 2.5\,\Omega} = 129.6\ W.$$

Example 4-3

A voltage source has a constant EMF of 28 volts and an internal resistance of $3\,\Omega$. It is connected to a load, R_L, which is in series with an additional resistance of $5\,\Omega$. What value of R_L will allow maximum power transfer to the load? Calculate the load power if this value of R_L is (a) doubled and (b) halved.

Solution

For maximum power transfer the value of R_L must be matched to the total of all the resistances not associated with the load. Therefore R_L must equal $5\,\Omega + 3\,\Omega = 8\,\Omega$.

$$\text{Maximum load power} = \frac{E^2}{4\,R_L} = \frac{(26\,V)^2}{4 \times 8\,\Omega} = 24.5\ W.$$

(a) If R_L is doubled to $16\,\Omega$.

$$\text{Load power} = \frac{(28\,V)^2 \times 16\,\Omega}{(3\,\Omega + 5\,\Omega + 16\,\Omega)^2} = 21.8\ W.$$

(b) If R_L is halved to $4\,\Omega$,

$$\text{Load power} = \frac{(28\,V)^2 \times 4\,\Omega}{(3\,\Omega + 5\,\Omega + 4\,\Omega)^2} = 21.8\ W.$$

VOLTAGE SOURCES IN SERIES

When voltage sources are connected in series-aiding as in Fig. 4-6, the total constant voltage EMF, E_T, is the sum of the individual EMFs. Therefore the normal purpose of connecting cells in series is to increase the voltage available. However the internal resistances are also additive so that the greater the number of voltage sources, the worse is the voltage regulation; this is the real disadvantage of the series connection. If N sources are joined in series-aiding:

$$E_T = E_1 + E_2 + E_3 + \text{------} + E_N.$$

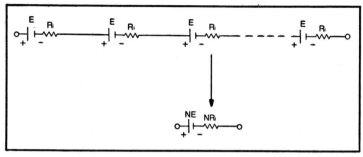

Fig. 4-6. Identical cells in series-aiding.

and $R_{iT} = R_{i2} + R_{i3} + \text{-----} + R_{iN}$. If each source has a constant voltage EMF, E, and an internal resistance R_i;

$$E_T = NE \text{ and } R_{iT} = NR_i.$$

Example 4-4

Figure 4-6 represents eight voltage sources each with an EMF of 1.5 V and an internal resistance of 0.1Ω. What load resistance may be connected across the sources for maximum power transfer? Calculate the value of the maximum load power.

Solution

Total EMF, $E_T = 8 \times 1.5\,\text{V} = 12\,\text{V}$.
Total internal resistance, $R_{iT} = 8 \times 0.1\,\Omega = 0.8\,\Omega$.
For maximum power transfer, $R_L = R_{iT} = 0.8\,\Omega$.

$$\text{Maximum load power} = \frac{E_T{}^2}{4\,R_L} = \frac{(12\,\text{V})^2}{4 \times 0.8\,\Omega} = 45\,W.$$

CONSTANT CURRENT SOURCES

The fact that the load voltage falls as the load current increases was explained by assuming that the source contained a constant voltage EMF in series with an internal resistance. However, this is only one possible assumption. The alternative is to consider that the source is a constant current generator which produces a variable EMF and has the same value of internal resistance in parallel. The generator current is equal to the short circuit value as already discussed.

The equivalence of the constant voltage and constant current generators is best illustrated by an example in which the open circuit voltage is 24 V and the short circuit current is 6 A. The internal resistance is therefore: 24V/6A = 4Ω. The circuits for the constant voltage and constant current generators are shown in Figs. 4-7 A and B. As a convention, the arrow in the symbol for the current generator indicates the direction of the electron flow.

Let a load of 8 Ω be connected across each of the generators. In the constant voltage circuit $V_L = 24V \times \dfrac{8\,\Omega}{8\,\Omega + 4\,\Omega} = 16$ V (voltage division rule) and therefore $I_L = 16V/8 \quad \Omega = 2A$. For the constant current generator $I_L = 6A \times \dfrac{4\,\Omega}{8\,\Omega + 4\,\Omega}$ = 2A (current division rule) and $V_L = 2A \times 8\,\Omega = 16$ V. Therefore, as far as the external load is concerned, the concepts of constant voltage and constant current generators are equally valid. However, the generator circuits themselves are not exact equivalents since under open-circuit load conditions, there is power dissipated in the internal resistance of the current generator but not in the internal resistance of the voltage generator.

The equations for the generators are:

1. *Constant Voltage Generator.* Constant voltage EMF = Open circuit terminal voltage.

$$\text{Internal resistance} = \frac{\text{Open circuit terminal voltage}}{\text{Short circuit terminal current}}$$

2. *Constant Current Generator.* Constant current = Short circuit terminal current.

$$\text{Internal resistance} = \frac{\text{Open circuit terminal voltage}}{\text{Short circuit terminal current}}$$

Example 4-5

A DC source has an open circuit terminal voltage of 100 V and a short circuit current of 20 A. A 20Ω load is connected across the terminals. Derive the values in the constant voltage and constant current equivalent circuits and in each case calculate the load current. What are the values of V_L, P_L and the percentage efficiency?

Solution

Constant Voltage Generator. Constant voltage EMF = 100 V. Internal resistance = 100 V/20 A = 5Ω.

$$\text{Load voltage, } V_L = \frac{100\,V \times 20\,\Omega}{20\,\Omega + 5\,\Omega} = 80\,V.$$

$$\text{Load current, } I_L = \frac{80\,V}{20\,\Omega} = 4\,A.$$

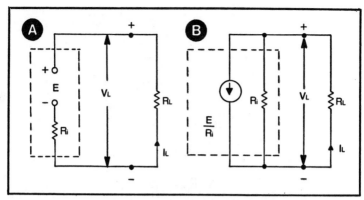

Fig. 4-7. Equivalent constant voltage and constant current generators.

Load power, $P_L = 80\,\text{V} \times 4\,\text{A} = 320\,W$.

Total power, $P_T = 100\,\text{V} \times 4\,\text{A} = 400\,W$..

Percentage efficiency $= \dfrac{P_L}{P_T} \times 100\% = \dfrac{320}{400} \times 100$

$= 80\%$

Constant Current Generator. Constant source current $= 20\,\text{A}$.

Load current, $I_L = 20\,\text{A} \times \dfrac{5\,\Omega}{20\,\Omega + 5\,\Omega} = 4\,A$.

CURRENT SOURCES IN PARALLEL

When combining two or more sources in parallel, it is inconvenient to use constant voltage equivalent circuits as shown in Fig. 4-8 A. The main difficulty is due to the circulating currents which exist between the sources so that at the very least the analysis will require the use of Kirchhoff's Laws. However, if each source is replaced by a constant current generator (Fig. 4-8 B), the current may simply be added to produce the total generator current; however if some of the sources are connected in parallel-opposing with respect to the load the currents must be summed algebraically. The total equivalent internal resistance is obtained by applying the reciprocal formula to the separate parallel resistances. This method of analyzing parallel sources is an application of Millman's Theorem (Chapter 8). If required, the final current generator may be transformed into its constant voltage equivalent.

If N sources are connected in parallel-aiding (all terminals of the same polarity are joined together).

$$I_T = I_1 + I_2 + I_3 + \text{----------} + I_N.$$

$$R_{iT} = \cfrac{1}{\cfrac{1}{R_{i1}} + \cfrac{1}{R_{i2}} + \cfrac{1}{R_{i3}} + \text{---} + \cfrac{1}{R_{iN}}}$$

If the N parallel sources are all identical,

$$I_T = NI$$

$$R_{iT} = \frac{R_i}{N}$$

where I and R_i are the constant current and equivalent resistance of one source. Total equivalent voltage, $E_T = I_T \times R_{iT} = NI \times R_i = I \times R_i$ = constant voltage EMF of one source (Fig. 4-9).

The advantage of connecting cells in parallel is therefore not to increase the voltage but to reduce the internal resistance. Normally equivalent constant voltage generators are used for sources in series while parallel sources are replaced by their equivalent constant current generators.

Example 4-6

In Fig. 4-8 A, the values are $E_1 = 10\,V$, $E_2 = 12\,V$, $E_3 = 8\,V$, $R_{i1} = 0.5\,\Omega$, $R_{i2} = 0.2\,\Omega$, $R_{i3} = 0.1\,\Omega$, $R_L = 2\,\Omega$. Calculate the value of V_L.

Solution

The values of the equivalent constant current generators are $I_1 = 10\,V/0.5\Omega = 20\,A$, $I_2 = 12\,V/0.2\,\Omega = 60\,A$, $I_3 = 8\,V/0.1\,\Omega = 80\,A$. The total equivalent current $= 20\,A + 60\,A + 80\,A = 160\,A$. The total internal resistance is given by:

$$R_{iT} = \cfrac{1}{\cfrac{1}{0.5} + \cfrac{1}{0.2} + \cfrac{1}{0.1}} = \frac{1}{2 + 5 + 10} = \frac{1}{17} = 0.0588\,\Omega.$$

Constant voltage EMF $= 0.0588\,\Omega \times 160\ \ A = 9.408\ \ V$. By the voltage division rule:

$$V_L = 9.408\,V \times \frac{2.0\,\Omega}{2.0\,\Omega + 0.0588\,\Omega} = 9.1\,V \text{ rounded off.}$$

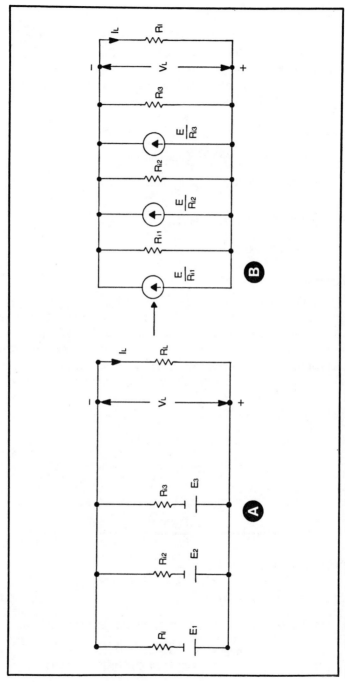

Fig. 4-8. Constant voltage and constant current equivalent circuits for cells in parallel.

75

Example 4-7

Figure 4-9 represents eight sources which are connected in parallel aiding. Each source has an EMF of 1.5 V and an internal resistance of 0.1Ω. What load resistance may be connected across the sources for maximum power transfer? Calculate the value of the maximum load power.

Solution

Total EMF, E_T = EMF of one source = 1.5 V

Total internal resistance, R_{iT} $\dfrac{0.1\Omega}{8}$ = $0.0125\,\Omega$.

For maximum power transfer, $R_L = R_{it}$ = $0.0125\,\Omega$.

Maximum load power = $\dfrac{(1.5\,V)^2}{4 \times 0.0125\,\Omega}$ = $45\,W$.

Example 4-8

Eight cells each with an EMF of 1.5 V and an internal resistance of 0.1Ω are connected in series aiding. Eight such banks are then connected in parallel aiding for a total of 64 cells. With such an arrangement what is the total EMF and the total internal resistance? What value of load resistance will provide maximum power transfer and what is the value of the maximum load power?

Solution

Total EMF, $E_T = 8 \times 1.5\,V = 12\,V$.

Total internal resistance, $R_{iT} = \dfrac{8 \times 0.1\Omega}{8}$ = 0.1Ω.

For maximum power transfer, $R_L = R_{iT}$ = $0.1\,\Omega$.

Maximum load power = $\dfrac{(12\,V)^2}{4 \times 0.1}$ = $360\,W$.

CHAPTER SUMMARY

☐ *Constant Voltage Source.*

Constant voltage EMF, E = Open-circuit terminal voltage.

Internal resistance, $R_i = \dfrac{\text{Open-circuit terminal voltage, E}}{\text{Short-circuit terminal current, } E/R_i}$.

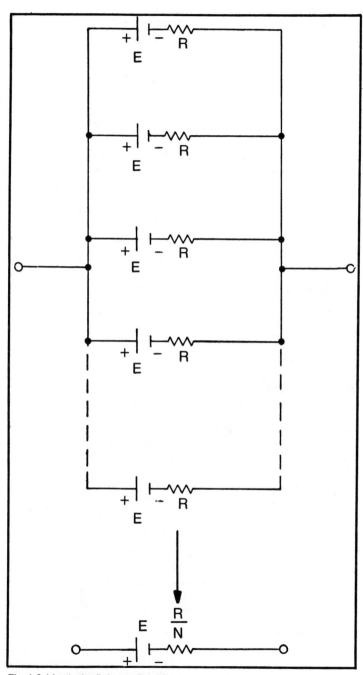

Fig. 4-9. Identical cells in parallel aiding.

Load current, $I_L = \dfrac{V_L}{R_L} = \dfrac{V_i}{R_i} = \dfrac{E}{R_L + R_i}$

Load voltage, $V_L = I_L R_L = \dfrac{E \times R_L}{R_L + R_i} = E - I_L R_i$

Percentage voltage regulation $= \dfrac{V_{NL} - V_{FL}}{V_{FL}} \times 100\%.$

Load power, $P_L = I_L{}^2 R_L = \dfrac{V_L^2}{R_L} = I_L V_L = \dfrac{E^2 R_L}{(R_L + R_i)^2}$

For maximum power transfer, $R_L = R_i$ (matching).

Maximum Load Power, $P_{Lmax} = \dfrac{E^2}{4R_L} = \dfrac{E^2}{4R_i}$

Percentage efficiency $= \dfrac{P_L}{P_T} \times 100\%.$

☐ *Sources in Series-Aiding*

$E_T = E_1 + E_2 + E_3 + \text{-----} + E_N$
$R_{iT} = R_{i1} + R_{i2} + R_{i3} + \text{----------} + R_{iN}$

For identical sources

$E_T = NE$ and $R_{iT} = NR_i.$

☐ *Constant Current Source.*

Constant current, I = Short-circuit terminal current.
$= E/R_i$

Load current, $I_L = I \times \dfrac{R_i}{R_L + R_i}$

☐ *Sources in Parallel Aiding.*

$I_T = I_1 + I_2 + I_3 + \text{-------} + I_N$

$R_{iT} = \dfrac{1}{\dfrac{1}{R_{i1}} + \dfrac{1}{R_{i2}} + \dfrac{1}{R_{i3}} + \text{---} + \dfrac{1}{R_{iN}}}$

For identical sources.

$I_T = NI \qquad R_{iT} = \dfrac{R_i}{N}$
$E_T = E$ (open-circuit voltage of one cell).

Magnetic Circuits

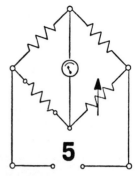

5

The total number of lines in a magnetic field is called the magnetic flux. The lines form a complete closed path which is referred to as a magnetic circuit. Figure 5-1 shows a simple magnetic circuit in which the iron ring provides the path for the magnetic flux. The mean length of the magnetic path is L meters and cross-sectional area of the ring is A square meters.

MAGNETIC FLUX

The S.I. unit of magnetic flux is the *Weber* (Wilhelm Eduard Weber, 1804-91) and its letter symbol is the Greek letter, ϕ (phi). The Weber (Wb) is defined in terms of *Faraday's Law of Electromagnetic Induction*; this law may be stated as follows:

When a conductor cuts or is cut by a magnetic flux, an EMF is induced in the conductor. The direction of the induced EMF depends on the direction of the magnetic field and on the direction in which the field moves relative to the conductor. The magnitude of the EMF is proportional to the rate at which the conductor cuts or is cut by the magnetic flux. Therefore the Induced EMF, E (volts)

$$= \frac{\text{Flux}, \phi \text{ (Webers)};}{\text{Time}, t \text{ (seconds)}}$$

and the weber is the magnetic flux which, when cut by a conductor in one second, generates an induced EMF of one volt. On the electromagnetic cgs system the flux unit is the line or Maxwell where 1 Weber = 10^8 Maxwells.

FLUX DENSITY

In Fig. 5-1 the magnetic flux has a cross-sectional area of A square meters; the flux density, B, is therefore defined by:

Flux density, B (Webers per square meter)

$$= \frac{\text{Flux (Webers)}}{\text{Cross-sectional Area, A (square meters).}}$$

The SI unit of flux density is the Tesla (T) so that 1 T = 1 Wb/m². In the cgs system the flux density unit is the Gauss (Maxwell/cm²) which is equivalent 1×10^{-4} tesla.

Faraday's Law relates to the generator effect in which mechanical energy is converted into electrical energy. The reverse action is the *motor effect* (Fig. 5-2) in which a force is exerted on a current-carrying conductor, situation in a magnetic field. In SI units: Force (Newtons) = Flux Density, B (Teslas) × current, I (amperes) × conductor's length, L (meters).

Referring again to Faraday's Law (Fig. 5-3) the induced EMF is: E (volts) = Flux density, B (teslas)

× conductor's length, L (meters)

× conductor's velocity, v (meters per second).

MAGNETO-MOTIVE FORCE

The magnetic flux in the iron ring of Fig. 5-1 is established as the result of the current, I, flowing through the exciting coil consisting of N turns. Increasing the current and the number of turns will increase the flux. This is equivalent to increasing the EMF applied across a resistor and observing that the current increases. The product of the current and the number of turns is therefore called the *magnetomotive* force (MMF), F. The SI unit of the MMF is the ampere-turn or ampere since the number of turns has no dimensions. Therefore: Magnetomotive Force, F (am-

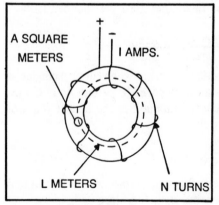

Fig. 5-1. Toroidal ring magnetic circuit.

A SQUARE METERS

I AMPS.

L METERS

N TURNS

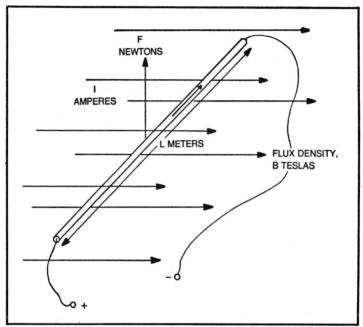

Fig. 5-2. Motor effect.

peres) = Current, I (amperes) × number of turns, N. The cgs unit of MMF is the Gilbert where 1 ampere-turn = $4\pi/10$ gilberts.

MAGNETIC FIELD INTENSITY

Using the example of the iron ring the MMF is distributed over a magnetic path whose mean length is L meters. The MMF per meter length of the magnetic circuit is called the magnetic field intensity or magnetizing force, H. The SI unit of magnetic field intensity is the ampere-turn per meter or ampere per meter so that:

Magnetic Field Intensity, H (amperes per meter)
$$= \frac{\text{Current, I (amperes)} \times \text{Number of turns, N}}{\text{Length, L (meters)}}.$$

In the cgs system the unit of magnetic field intensity is the Oersted where 1 ampere-turn per meter = $4\pi \times 10^{-3}$ oersted.

PERMEABILITY OF FREE SPACE

In Fig. 5-4, X represents the cross-section of a long straight conductor which is positioned in a vacuum and carries an electron current of one ampere emerging from the paper. The magnetic flux

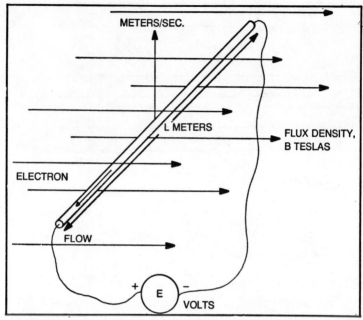

Fig. 5-3. Faraday's law: generator effect.

surrounding X will be in the shape of concentric circles. One such circle, marked C, has a radius of 1 meter and represents the path of one line of flux. The MMF acting on this path is one ampere and since the path length is 2π meters, the magnetic field intensity is $\frac{1}{2}\pi$ ampere per meter.

If B is the flux density in teslas around the circle, C, and Y is a second conductor which is parallel to X and also carries a current of one ampere, then by the equation: Force exerted on Y = B (Teslas) × 1 (Ampere) × 1 (Meter).

But, from the definition of the ampere in Chapter 2, this force must be 2×10^{-7} newton. Therefore B=2×10^{-7} telsa.

$$\frac{\text{Flux Density at Y, B}}{\text{Magnetic Field Intensity at Y, H}} \quad = \quad \frac{2 \times 10^{-7} \text{ T}}{\frac{1}{2}\pi \text{ A/m}}$$

$$= \quad 4\pi \times 10^{-7}$$

$$= \quad 12.57 \times 10^{-7} \text{ SI units.}$$

The ratio of $\frac{B}{H}$ for a vacuum is called the permeability of free space whose letter symbol is μ_o; consequently B = μ_oH. The value of μ_0 is determined by the system of units; for example in the cgs system the value of μ_0 is 1.

Many substances are nonmagnetic and have the same permeability as free space. However there are a few paramagnetic materials such as aluminum and platinum which have a permeability slightly exceeding that of free space. Yet other elements, for example copper and silver are diamagnetic and have a permeability slightly less than free space. Finally the ferromagnetic substances (iron, nickel, cobalt) have permeabililities many times greater than free space.

Example 5-1

A coil of 500 turns carries a current of 6 A and is wound uniformly over a nonmagnetic toroidal ring with a mean circumference of 40 cm and a cross-sectional area of 3 cm². Calculate the magnetic field intensity, the flux density and the total flux.

Solution

Magnetomotive force, $F = 6\,A \times 500 = 3000$ ampere-turns.

Length of magnetic path, $L = 40\,cm = 0.4\,m$.

Magnetic field intensity, $H = \dfrac{F}{L} = \dfrac{3000}{0.4} = 7500$ amperes/meter.

Flux density, $B = \mu_0 H$.

$$= 12.57 \times 10^{-7} \times 7500$$

$$= 0.00943\ tesla$$

Cross-sectional area, $A = 3\,cm^2$
$$= 3 \times 10^{-4} m^2.$$

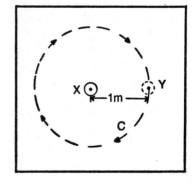

Fig. 5-4. Two parallel conductors in a vacuum.

Total flux, $\phi = B \times A$
$$= 0.00943T \times 3 \times 10^{-4} m^2$$
$$= 2.83 \ \mu Wb.$$

Example 5-2

An airgap is 5 mm long and has a cross sectional area of 30 cm². What is the number of ampere-turns required to establish a total flux of 8.0 mWb in the airgap?

Solution

Cross-sectional area, $A = 30 \times 10^{-4} m^2$.

Flux Density, $B = \dfrac{\phi}{A} = \dfrac{8.0 \times 10^{-3}}{30 \times 10^{-4}} = 2.67 \ T.$

Magnetic Field Intensity, $H = \dfrac{B}{\mu_o} = \dfrac{2.67}{12.57 \times 10^{-7}},$

$$= 2.124 \times 10^6$$
ampere-turns.

Length of magnetic path, $L = 5 \times 10^{-3} m.$

Total ampere-turns $= 2.124 \times 10^6 \times 5 \times 10^{-3}$
$$= 10620.$$

RELATIVE PERMEABILITY

It is well known that the nonmagnetic flux inside a coil is intensified when a soft iron core is inserted. Consequently, if the magnetic core of the toroidal ring in Fig. 5-1 is replaced by an iron core, the flux for a given MMF is greatly increased. The ratio of the flux density established in the core material to the flux density produced in a vacuum (or nonmagnetic core) for the same magnetic field intensity, is called the relative permeability, μr. For air $\mu r = 1$ but for certain nickel-iron alloys, the relative permeability may be as high as 100,000. It should be emphasized that the value of μr depends on the magnitude of the magnetic field intensity and is not a constant for a particular alloy (Fig. 5-5).

For a core material with a relative permeability, μr
$\quad$ $B = \mu o \, \mu r \, H.$

Absolute permeability, $\mu = \dfrac{B}{H} = \mu o \, \mu r.$

Referring back to Fig. 5-1 the iron ring has a cross-sectional area of A square meters and a mean circumference of L meters. It is wound with N turns of a coil in which a current of I amperes flows.

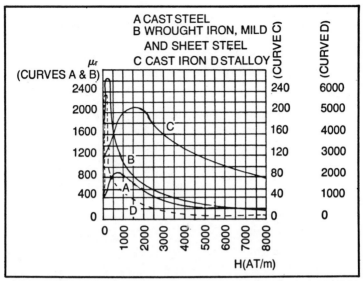

Fig. 5-5. Relationship between relative permeability and magnetic field intensity.

Then: Total flux, ϕ = Flux density, B × Area, A. Magnetomotive Force, F = Magnetic Field Intensity, H × length, L.

$$\frac{\text{MMF, F}}{\text{Flux,}\phi} = \frac{\text{HL}}{\text{BA}} = \frac{\text{HL}}{\mu_0\,\mu r\,\text{HA}} = \frac{\text{L}}{\mu_0\,\mu r\,\text{A}}$$

This relationship is sometimes known as Rowland's Law which is the magnetic equivalent of the Ohm's Law electrical equation.

$$\frac{\text{EMF, E}}{\text{Current, I}} = \text{Resistance, R} = \frac{\text{resistivity, }\rho\times\text{length, L}}{\text{cross-sectional area, A}}$$

The quantity

$$\frac{\text{L}}{\mu_o\,\mu_r\,\text{A}}$$

is therefore comparable with resistance and is called the reluctance, R, of the magnetic circuit. There are no special SI units for reluctance which can be referred to as ampere-turns per weber or as the reciprocal of the henry which is the unit of inductance (Chapter 6). The reciprocal of reluctance is the permeance, P, whose SI unit is the henry or weber per ampere-turn.

On the cgs system the unit of reluctance is the rel which is equivalent to one gilbert per maxwell; 1 SI unit of reluctance = $4\pi \times 10^{-9}$ rels.

85

When comparing $R = \dfrac{\rho L}{A} = \dfrac{L}{\sigma A}$ and $R = \dfrac{L}{\mu_0 \mu_r A} = \dfrac{L}{\mu A}$,

the absolute permeability μ, corresponds to the conductivity, σ; the SI unit of permeability is the henry per meter.

Example 5-3

A particular magnetic circuit has an air gap which is 7 mm long and has a cross-sectional area of 30 cm^2. Calculate the reluctance of the gap and the MMF required to establish a flux of 900 μWb across the gap.

Solution

$$\text{Reluctance of the gap, } R = \frac{L}{\mu_0 A} = \frac{7 \times 10^{-3} \text{m}}{12.57 \times 10^{-7} \times 30 \times 10^{-4} \text{m}^2}$$

$$= 1.86 \times 10^{6} \text{ ampere-turns per weber.}$$

$$\text{Required MMF} = R \times \phi = 1.86 \times 10^6 \times 900 \times 10^{-6},$$

$$= 1670 \text{ ampere-turns.}$$

Example 5-4

An iron ring has a mean circumference of 60 cm and a uniform cross-sectional area of 5 cm^2. Corresponding to a flux density of 1.4 Wb/m^2, the relative permeability of the iron is 2400. For the value of this flux density, calculate the reluctance of the circuit and the MMF required.

Solution

$$\text{Reluctance of the iron ring, } R = \frac{L}{\mu_0 \mu_r A}$$

$$= \frac{60 \times 10^{-2}}{12.57 \times 10^{-7} \times 2400 \times 5 \times 10^{-4}}$$

$$= 3.98 \times 10^5 \text{ ampere-turns per weber}$$

$$\text{Flux, } \phi = BA = 1.4 \times 5 \times 10^{-4} = 7 \times 10^{-4} \text{ Wb.}$$

$$\text{Required MMF} = R \times \phi = 3.98 \times 10^5 \times 7 \times 10^{-4} = 280 \text{ ampere-turns}$$

SERIES MAGNETIC CIRCUIT

Figure 5-6 shows a composite magnetic circuit consisting of three specimens of iron and an air gap. It is assumed that the same

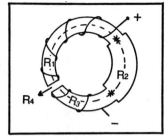

Fig. 5-6. Composite series magnetic circuit.

flux is established throughout the circuit and that the magnetic leakage is negligible. The principles used in analyzing a series electrical circuit can then be applied to the composite magnetic circuit. Therefore: Total Reluctance, $R_T = R_1 + R_2 + R_3 + R_4$ Total MMF, $F_T = F_1 + F_2 + F_3 + F_4$ There are comparable similarities between the equations for parallel electrical and magnetic circuits (Example 5-7).

One important difference between electrical and magnetic circuits lies in the concept of power. While energy must be continuously supplied to an electrical circuit to sustain the flow of electricity, the magnetic flux, once set up, does not need any further energy. In the case of the magnetized iron ring the only energy supplied after the flux is fully established, is dissipated by the coil's resistance in the form of heat.

MAGNETIZATION CURVE

Since the reluctance for a particular type of iron depends on the flux density and is not a constant, it is common practice to use the magnetization curve to obtain directly the corresponding values of the flux density, B, and the magnetic field intensity, H. This curve is derived experimentally and illustrates the relationship between B and H for a given specimen of iron; examples are shown in Fig. 5-7A. With each curve there is an initial sharp rise after which there is a levelling off due to saturation of the iron.

In Fig. 5-7B OAC represents the magnetization curve for a certain type of iron. If after reaching a maximum value corresponding to OK, the magnetic field intensity is reduced (by decreasing the current in the exciting coil), the flux density follows the curve CD. When the field intensity, H, is reduced to zero, the value of B still present in the iron is represented by OD which is called the residual or remanent flux density. If the current is now reversed so that the field intensity is increased in the opposite direction, there will be a value OE for which the magnetic flux

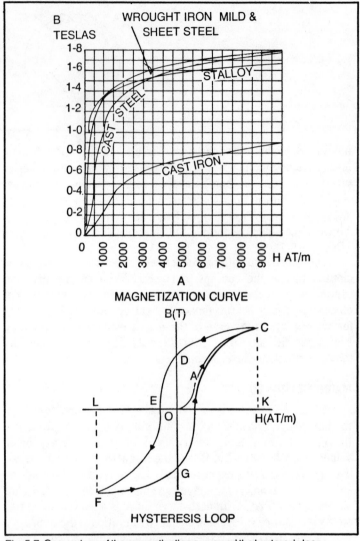

Fig. 5-7. Comparison of the magnetization curve and the hysteresis loop.

density is reduced to zero; this value of H is called the coercive force. Further increasing the field intensity will raise the flux density in the reverse direction (as represented by the curve EF).

HYSTERESIS LOOP

If the field intensity is now changed from OL back to OK, the flux density follows the curve FGC which is similar to the curve

CDEF. The complete closed figure is called the hysteresis loop; the word *hysteresis* indicates that the result (the flux density, B) lags behind the cause (the field intensity, H). Hysteresis produces a heating effect in the iron and therefore represents an energy loss from the source of the exciting coil. The amount of energy loss in joules per cubic meter is represented by the area of the loop. This is indicated in the Steinmetz equation:

Hysteresis power loss $= kfvB_{max}^{1.6}$ watts

> where k = a constant for a particular specimen of iron.
> f = frequency in hertz.
> v = volume of iron in cubic meters.
> B_{max} = maximum flux density in teslas corresponding to the peak field intensity in ampere turns per meter.

For most ferromagnetic materials the ratio of the residual flux density to the maximum flux density is about 0.7. However the coercive force can range from more than 10,000 to less than 10 ampere-turns per meter.

The easiest way to demagnetize the iron specimen and return to the point 0, is to reverse the exciting current many times while gradually reducing the value of the peak current to zero.

Example 5-5

A cast-iron ring has a mean circumference of 50 cm and a cross-sectional area of 4 cm². What is the current required in a coil of 300 turns to establish a total flux of 160 μWb?

Solution

Flux density, $B = \dfrac{\phi}{A} = \dfrac{160 \times 10^{-6}}{4 \times 10^{-4}} = 0.4$ Wb/m².

Using Fig. 5-7A, the magnetic field intensity, H, corresponding to a flux density of 0.4 Wb/m² in cast-iron, is 1500 ampere-turns per meter.

Required MMF, $F = H \times L = 1500 = 50 \times 10^{-2}$
$= 750$ ampere-turns.

Magnetizing current $= \dfrac{750}{300} = 2.5A$.

Example 5-6

Figure 5-8 represents a cast steel ring with a cross-sectional area of 6 cm² and a mean circumference of 55 cm. There are two

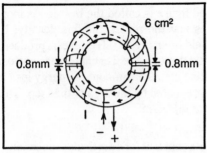

Fig. 5-8. Diagram for example 5-6.

radial cuts at diametrically opposite points and these are each filled with nonmagnetic material to a thickness of 0.8 mm. Assuming no magnetic leakage, calculate the total MMF required to establish a flux density of 1.5 T.

Solution

From Fig. 5-7A the number of ampere-turns per meter required for the cast steel is 3000. The corresponding number of ampere-turns is $3000 \times 55 \times 10^{-2} = 1650$.

For the nonmagnetic gaps the magnetic field intensity is:

$$H = \frac{1.5}{12.57 \times 10^{-7}} = 1.19 \times 10^6 \text{ ampere-turns per meter.}$$

Total ampere-turns for the two gaps $= 2 \times 1.19 \times 10^6 \times 0.8 \times 10^{-3} = 1900$

Total MMF required $= 1650 + 1900 = 3550$ ampere-turns.

Example 5-7

Figure 5-9 shows a cast iron magnetic core such as used in transformers. The cross-sectional area of each side limb, is 10 cm² while the center limb's area is 13 cm². Neglecting any magnetic leakage calculate the MMF required to establish a flux density of 0.2 T in the airgap.

Solution

This magnetic arrangement is equivalent to a series-parallel circuit. The right side limb is in series with the airgap and their total reluctance is in parallel with the left side limb. Finally this combination is in series with the center limb.

From Fig. 5-7A a flux density of 0.2 T in cast iron is produced by a magnetic field intensity of 800 ampere-turns per meter.

Therefore the ampere turns for the iron in the right hand limb is 800 × 0.3 = 240. The magnetic field intensity for the airgap is

$$\frac{B}{\mu_o} = \frac{0.2}{12.57 \times 10^{-7}} = 1.6 \times 10^5$$

and the corresponding ampere turns are

$$1.6 \times 10^5 \times 0.1 \times 10^{-2} = 160.$$

The total ampere turns for the left hand limb will be

$$240 + 160 = 400$$

which will result in a magnetic field intensity of

$$\frac{400}{30 \times 10^{-2}} = 1333 \text{ ampere turns per meter.}$$

Again using Fig. 5-7A the corresponding flux density is 0.375 T. Therefore the total flux in the center limb is

$$(0.2 + 0.375) \times 10 \times 10^{-4} = 5.75 \times 10^{-4} \text{ Wb,}$$

and the flux density is $\dfrac{5.75 \times 10^{-4}}{13 \times 10^{-4}} = 0.44$ T

which is produced by a magnetic field intensity of 1750 ampere-turns per meter. The ampere-turns for the center limb are 1750 × 20 × 10^{-2} = 350, and the total MMF required is 400 + 350 = 750 ample turns.

Table 5-1 shows a comparison between electric and magnetic units.

CHAPTER SUMMARY

☐ **Equations**

MMF, $F = IN$ ampere-turns.

Magnetic field intensity, $H = \dfrac{IN}{L}$ ampere-turns/m

Flux density, $B = \dfrac{\phi}{A}$ tesla or WB/m².

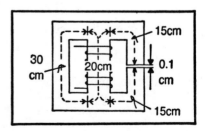

Fig. 5-9. Diagram for example 5-7.

Table 5-1. Comparison between Electric and Magnetic Units.

ELECTRIC CIRCUIT OHM'S LAW, RESISTANCE =EMF/Current		MAGNETIC CIRCUIT Rowland's Law, Reluctance =MMF/Flux	
QUANTITY	SI UNIT	QUANTITY	SI UNIT
EMF	Volt	MMF	ampere-turn
Current	ampere	Flux	weber
Current density	ampere per square meter	flux density	weber per square meter or telsa
Resistance	Ohm	Reluctance	ampere-turn per weber
Conductance	siemens	Permeance	henry or weber per ampere-turn
Conductivity	siemens per meter	permeability	henry per meter
Electric field intensity	volts per meter	Magnetic field intensity	ampere-turns or amperes per meter

Permeability of free space $= 4\pi \times 10^{-7} = 12.57 \times 10^{-7}$ SI units.

Absolute permeability, $\mu = \mu_o \mu_r$.

For nonmagnetic materials, $B = \mu_o H$.

For magnetic materials, $B = \mu_o \mu_r H$.

Reluctance, $R = \dfrac{L}{\mu A} = \dfrac{L}{\mu_o \mu_r A}$ reciprocal henry.

Permeance, $P = \dfrac{1}{R} = \dfrac{\mu A}{L}$ henry

☐ **Series magnetic circuit**

$R_t = R_1 + R_2 + R_3 + \text{-----} + R_N$.

☐ **Parallel magnetic circuit**

$$\frac{1}{R_T} = \frac{1}{R_1} + \frac{1}{R_2} + \frac{1}{R_3} + \text{-----} + \frac{1}{R_N}$$

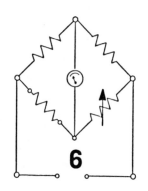

Inductance

6

Inductance is that property of an electrical circuit which opposes any *sudden change* of current. This compares with the property of resistance which only limits or opposes the flow of current. The circuit of Fig. 6-1 is used to illustrate the differences between the two properties. Although a straight conductor possesses inductance, the property is most marked in a coil which is referred to as an inductor with L as the letter symbol.

INTRODUCTION TO INDUCTANCE

As soon as the switch, S, is closed, the current in the resistor immediately jumps from zero to a value of E/R amperes. The potential difference or voltage drop across R exactly balances (opposes) the applied voltage, E. It is therefore appropriate to write $V_R = -I_r \times R$.   By Kirchhoff's Voltage Law $E + V_R = 0$ which leads to $E = I_R \times R$.

By contrast, as soon as the current starts to grow in the coil, a magnetic flux is created and then, as this expands out, it cuts the turns of the coil. By Faraday's Law a counter EMF, V_L, is self-induced into the coil and this voltage must exactly balance (oppose) the applied EMF.

Since the counter EMF is produced by a moving flux, this requires a changing current (a constant flux would generate zero induced voltage). The current therefore starts at zero and has a rate of growth which would be measured in amperes per second. Summarizing, the current in the coil is zero before the switch is closed and remains at zero just after the switch is closed; in other words, inductance has opposed any sudden change in the current's value.

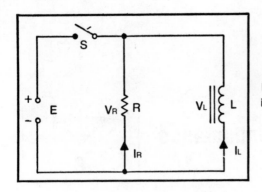

Fig. 6-1. Resistance and inductance in a dc circuit.

The size of the counter EMF is determined by the rate of cutting of the magnetic flux. This in turn will depend on the rate of change of current and some factor related to the coil. This factor is the inductance which is associated with the coil's number of turns, nature of the core, cross-sectional area and length. In equation form:

$$V_L = -L \times \frac{\Delta^I L}{\Delta^T}$$

V_L = the counter EMF with the negative sign indicating that V_L opposes E.

$\dfrac{\Delta^I L}{\Delta T}$ = the rate of change of current in amperes per second.

The symbol "Δ" means "the change in" so that ΔI_L is the change in current and ΔT the corresponding change in time. L = the coil's inductance measured in henrys (H).

The inductance is one henry (Joseph Henry, 1797 - 1878) if when the current is changing at the rate of one ampere per second, the induced EMF is one volt. By Kirchhoff's Voltage Law:

$$E + V_L = 0.$$

Then:

$$E = L \times \frac{\Delta^I L}{\Delta T}$$

or: Rate of change of current

$$\frac{\Delta^I L}{\Delta T} = \frac{E}{L}$$

amperes per second.

94

The current in the inductor therefore starts at zero and grows at a constant rate. However, this neglects the coil's resistance, R_L. In practice, I_L takes an appreciable time to reach a final steady value equal to E/R_L amperes. When this final condition has been reached, the flux is stationary and the counter EMF is zero. If S is now opened, there is no immediate change in the inductor's current. However, as I_L subsequently begins to decay, the magnetic flux starts to collapse and induces a counter EMF whose polarity is such as to oppose the decrease in I_L. This is an example of *Lenz's Law* which states that the polarity of the induced EMF is such as to oppose the change producing the EMF. The return path for I_L is through the resistor so that the current through both components is now the same and will again take an appreciable time to decrease to zero.

Example 6-1

The current through an inductor changes from 7A to 2A in a time of 25 milliseconds. If the self-inductance is 1.5 henry, what is the average value of the induced EMF?

Solution

Change in current, $\Delta I = 7A - 2A = 5A$.
Corresponding change in time, $\Delta T = 25 \times 10^{-3}$ seconds.

$$\text{Average induced EMF} = L \times \frac{\Delta I}{\Delta T} = 1.5H \times \frac{5A}{25 \times 10^{-3} \text{ sec.}}$$

$$= 300 \, volts.$$

FACTORS DETERMINING A COIL'S INDUCTANCE

Consider a coil of N turns with a length of ι meters and a cross-sectional area of A square meters (Fig. 6-2). The current in this coil changes by ΔI amperes in a corresponding time change of ΔT seconds. This coil is uniformly wound on a magnetic core with a relative permeability, μ_r. Using Faraday's Law and regarding each turn as a conductor

$$E = N \frac{\Delta \phi}{\Delta T} = L \frac{\Delta I}{\Delta T}$$

where $\Delta \phi$ in webers is the change in flux associated with the change, ΔI. Therefore:

$$L = N \frac{\Delta \phi}{\Delta I} = \frac{\text{change in flux-linkages}}{\text{change of current}}.$$

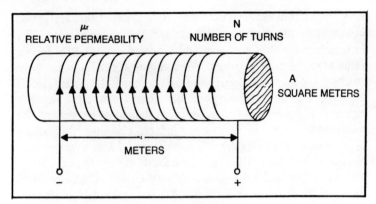

Fig. 6-2. Factors affecting the self-inductance of a coil.

The term *flux linkages* means the product of the flux and the number of turns through which the flux passes or with which the flux is linked. This leads to an alternative definition of the henry. The inductance of a coil is one henry if a current change of one ampere produces a change in flux-linkages of 1 weber-turn.

Therefore:
$$L = N\frac{\Delta\phi}{\Delta I}$$

$$= NA \times \frac{\Delta B}{\Delta I}$$

$$= NA\mu_o\mu_r \times \frac{\Delta H}{\Delta I}$$

$$= NA\mu_o\mu_r\frac{N}{\iota} \times \frac{\Delta I}{\Delta T}$$

$$= \frac{\mu_o\mu_r N^2 A}{\iota} = \frac{N^2}{R} \text{ henrys}$$

(R is the reluctance of the magnetic path.)

Since the inductance depends on the relative permeability, it will not be a constant but will vary with the current carried by the coil. The variation of inductance with current will be similar in shape to the μ_r/H curves of Fig. 5-5. For a coil with a nonmagnetic core, the inductance is: $L = \frac{\mu_o N^2 A}{\iota} = 12.57 \times 10^{-7} \times \frac{N^2 A}{\iota}$ henrys.

Example 6-2

A coil of 500 turns is wound on a nonmagnetic core. A current of 6 A flows through the coil and establishes a flux of 300 μWb.

What is the inductance of the coil and the average value of the induced EMF if the current is reversed in 25 milliseconds?

Solution

Inductance, $L = N \times \dfrac{\Delta\phi}{\Delta I}$

$$= 500 \times \dfrac{300 \times 10^{-6}}{6} \text{ H}$$

$$= 25\,mH.$$

Average induced EMF, $E = N \times \dfrac{\Delta\phi}{\Delta T}$

$$= 500 \times \dfrac{2 \times 300 \times 10^{-6}}{25 \times 10^{-3}} = 12 \text{ V}.$$

Alternatively, $E = L \times \dfrac{\Delta I}{\Delta T} = 25 \times 10^{-3} \times \dfrac{2 \times 6}{25 \times 10^{-3}}$

$$= 12 V.$$

Example 6-3

A cast iron ring has a mean circumference of 50 cm and a cross-sectional area of 4 cm². The ring is uniformly wound with a coil of 300 turns. Calculate the inductance if currents of (a) 2A and (b) 8A are reversed in the coil.

Solution

(a) Magnetic field intensity, $H = \dfrac{IN}{L} = \dfrac{2 \times 300}{50 \times 10^{-2}}$

$$= 1200 \text{ ampere-turns per meter}$$

From Fig. 5-7A: Flux density, $B = 0.3$T.

Flux, $\phi = BA = 0.3 \times 4 \times 10^{-4} \text{ Wb} = 120\,\mu\text{Wb}.$

Inductance, $L = N \times \dfrac{\Delta\phi}{\Delta I}$

$$= 300 \times \dfrac{2 \times 120 \times 10^{-6}}{2 \times 2} \text{ H}$$

$$= 0.018\,H.$$

(b) Magnetic field intensity, $H = \dfrac{8 \times 300}{50 \times 10^{-2}}$

$\qquad = 4800$ ampere-turns per meter

Corresponding flux density, $B = 0.71$ T.

Flux, $\phi = BA = 0.71 \times 4 \times 10^{-4}$ Wb $= 284\,\mu$ Wb.

Inductance, $L = 300 \times \dfrac{2 \times 284 \times 10^{-6}}{2 \times 8}$ H

$\qquad = 0.011$ H.

This example illustrates the fact that the value of the inductance depends on the current.

Example 6-4

If the cast iron ring of example 6-3 is replaced by a nonmagnetic core, calculate the new inductance of the coil.

Solution

New inductance, $L = \dfrac{\mu_0 N^2 A}{\ell}$

$$= \dfrac{12.57 \times 10^{-7} \times (300)^2 \times 4 \times 10^{-4}}{50 \times 10^2}\ \text{H}$$

$= 90.5\,\mu$H.

This shows that a coil with an inductance of this order of henrys will require some form of iron core.

ENERGY STORED IN THE MAGNETIC FIELD SURROUNDING AN INDUCTOR

The magnetic field associated with an inductor represents a form of energy which is created by the current flowing through the coil. The energy is delivered by the source from which the current is drawn. Assume that the coil has a constant inductance of L henrys and that after a switch is closed, the current grows at a constant rate from zero to I amperes in a time of T seconds (Fig. 6-3). The average current is $\dfrac{I}{2}$ and the source voltage, E, is $\dfrac{LI}{T}$.

Average power delivered to the magnetic field

$$= E \times \frac{I}{2} \quad = \frac{LI}{T} \quad \times \quad \frac{I}{2} \quad = \frac{LI^2}{2T}\ \text{watts}$$

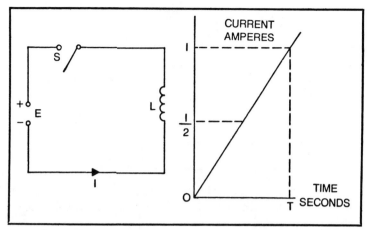

Fig. 6-3. Energy stored in an inductor.

Total energy stored in magnetic field = Average power
$$\times \text{ time}$$
$$= \frac{LI^2}{2T} \times T$$
$$= \frac{1}{2}LI^2 \text{ joules.}$$

If the switch is now opened, the current decays and the energy in the magnetic field is dissipated in the form of the arc which occurs between the contacts of the switch.

Example 6-5

A coil with an inductance of 4 H and a resistance of 25Ω is connected across a 60 V source. Calculate (a) the initial rate of current growth, (b) the value of the final current and (c) the energy stored in the magnetic field when the final conditions have been reached.

Solution

(a) Initial rate of growth of current $= \dfrac{E}{L}$

$$= \frac{60 \text{ V}}{4 \text{ H}}$$
$$= 15 \text{ amperes per second.}$$

(b) Final current $= \dfrac{E}{R} = \dfrac{60 \text{ V}}{25 \, \Omega} = 2.4 \, A.$

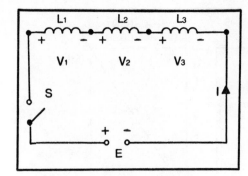

Fig. 6-4. Inductors in series.

There may be some difficulty in reconciling the answers to (a) and (b) until it is realized that the current reached its final value in less than one second.

(c) Energy stored = ½LI² = ½ × 4 × (2.4)² = *11.5 joules*.

INDUCTORS IN SERIES AND PARALLEL

Referring to Fig. 6-4 the inductors L_1, L_2 and L_3 are in series so that the same rate of change of current is associated with each inductor. Therefore:

$$V_1 = -L_1 \frac{\Delta I}{\Delta T} \ , \ V_2 = -L_2 \frac{\Delta I}{\Delta T}, \ V_3 = -L_3 \frac{\Delta I}{\Delta T}$$

and $E + V_1 + V_2 + V_3 = 0$.

This leads to: $E = \frac{\Delta I}{\Delta T} (L_1 + L_2 + L_3)$.

If L_T is the total equivalent inductance: $E = L_T \frac{\Delta I}{\Delta T}$ so that $L_T = L_1 + L_2 + L_3$. This is similar to the formula for resistors in series.

In Fig. 6-5 the inductors L_1, L_2, and L_3 are in parallel so that the total rate of change of current associated with the source is equal to the sum of the rates of change of current in the three branches. Then:

$$\frac{\Delta I_1}{\Delta T} = \frac{E}{L_1} \ , \ \ \frac{\Delta I_2}{\Delta T} = \frac{E}{L_2} , \ \frac{\Delta I_3}{\Delta T} = \frac{E}{L_3}$$

and: $\frac{\Delta I_T}{\Delta T} = \frac{\Delta I_1}{\Delta T} + \frac{\Delta I_2}{\Delta T} + \frac{\Delta I_3}{\Delta T} = E \times \left(\frac{1}{L_1} + \frac{1}{L_2} \times \frac{1}{L_3} \right)$

If L_T is the total equivalent inductance

$$\frac{\Delta I_T}{\Delta T} = \frac{E}{L_T} = E \times \left(\frac{1}{L_1} + \frac{1}{L_2} + \frac{1}{L_3}\right)$$

and: $\frac{1}{L_T} = \frac{1}{L_1} + \frac{1}{L_2} + \frac{1}{L_3}.$

The same type of reciprocal formula applies to resistors in parallel.

Example 6-6

In Fig. 6-4 $L_1 = 1.8$ H, $L_2 = 3.7$ H, $L_3 = 2.4$ H and $E = 38$ V. Calculate the voltages V_1, V_2, V_3 and the rate of growth of current. Determine the energy stored in each inductor after 5 seconds from the time S is closed.

Solution

Total inductance,

$$L_T = L_1 + L_2 + L_3 = 1.8 + 3.7 + 2.4 = 7.9 \text{ H.}$$

Rate of growth of current $= \dfrac{38\text{V}}{7.9\text{H}} = 4.81$ amperes per second.

$V_1 = 4.81 \times 1.8 = 8.66$ V.
$V_2 = 4.81 \times 3.7 = 17.80$ V.
$V_3 = 4.81 \times 2.4 = 11.54$ V.

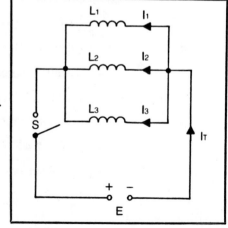

Fig. 6-5. Inductors in parallel.

Check: $V_1 + V_2 + V_3 = 8.66 + 17.80 + 11.54 = 38$ V.

After 5 seconds the circuit current, I, is $4.81 \times 5 = 24.05$ A.

Energy stored in the 1.8 H inductor $= \frac{1}{2} \times 1.8 \times (24.05)^2 = $ *521 joules*.

Energy stored in the 3.7 H inductor $= \frac{1}{2} \times 3.7 \times (24.05)^2 = $ *1070 joules*.

Energy stored in the 2.4 H inductor $= \frac{1}{2} \times 2.4 \times (24.05)^2 = $ *694 joules*.

Total energy stored in $521 + 1070 + 694 = 2285$ joules.

Check: Total energy stored $= \frac{1}{2} \times L_T \times I^2 = \frac{1}{2} \times 7.9 \times (24.05)^2 = 2285$ joules.

Example 6-7

In Fig. 6-5, $L_1 = 1.8$ H, $L_2 = 3.7$ H, $L_3 = 2.4$ H and E = 38 V. Calculate the rates of current growth in the individual inductors and the total rate of current growth. Determine the energy stored in each inductor after 5 seconds from the time S is closed.

Solution

Total inductance, L_T, is given by:

$$\frac{1}{L_T} = \frac{1}{1.8} + \frac{1}{3.7} + \frac{1}{2.4} = 0.555 + 0.270 + 0.417 = 1.242$$

$$L_T = \frac{1}{1.242} = 0.805 \, \text{H}.$$

Total rate of current growth is
$$\frac{38}{0.850} = 47.2 \text{ amperes per second.}$$
Rate of current growth in the 1.8 H inductor is
$$\frac{38}{1.8} = 21.1 \text{ amperes per second.}$$
Rate of current growth in the 3.7 H inductor is
$$\frac{38}{3.7} = 10.3 \text{ amperes per second.}$$
Rate of current growth in the 2.4 H inductor is
$$\frac{38}{2.4} = 15.8 \text{ amperes per second.}$$

Check: Total rate of current growth = $21.1 + 10.3 + 15.8$ = *47.2* amperes per second.

Fig. 6-6. RL time constant.

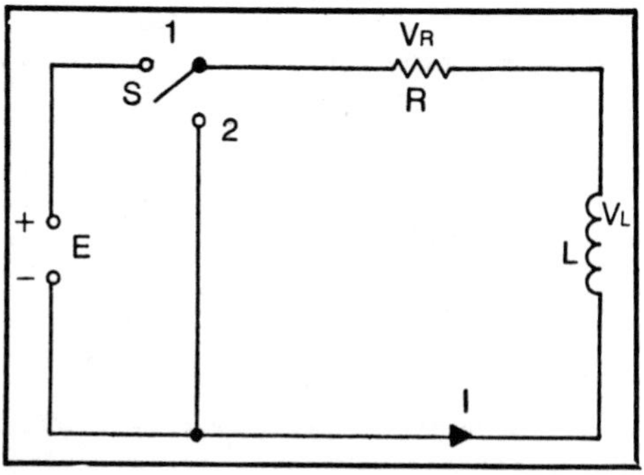

After 5 seconds the currents in the individual inductors are 21.1 × 5 = 105.5 A, 10.3 × 5 = 51.5 A, 15.8 × 5 = 79.0 A and the total current is 47.2 × 5 = 236.0 A.

Energy stored in the 1.8 H inductor is

$$\frac{1}{2} \times 1.8 \times (105.5)^2 = \textit{10017 joules.}$$

Energy stored in the 3.7 H inductor is

$$\frac{1}{2} \times 3.7 \times (51.5)^2 = \textit{4907 joules.}$$

Energy stored in the 2.4 H inductor is

$$\frac{1}{2} \times 2.4 \times (79.0)^2 = \textit{7489 joules.}$$

Total energy stored is 10017 + 4907 + 7489 = 22415 joules.

Check: Total energy stored is $\frac{1}{2} \times 0.805 \times (236)^2 = 22415$ joules.

TIME CONSTANT OF AN INDUCTOR AND A RESISTOR IN SERIES

Immediately after the switch S of Fig. 6-6 is closed in position 1, the inductance prevents any sudden change of current so that the initial circuit conditions are:

$V_L = E$, $V_R = 0$, $I = 0$ and the rate of growth of current is E/L amperes per second.

As the current increases, V_R rises and therefore V_L falls so that the rate of current growth is less.

The time constant of the circuit is L/R seconds where L is the inductance in henrys and R is the resistance in ohms. This is the time taken for the current to reach 63.2% of its final steady-state value which is only limited by the resistance, R. The order of the time constant's value may be estimated from the values shown in Table 6-1.

103

Table 6-1. LR Time Constants.

L	R	Time constant L/R
H	Ω	seconds
H	kΩ	milliseconds
H	MΩ	microseconds
mH	Ω	milliseconds
mH	kΩ	microseconds
mH	MΩ	nanoseconds
μH	Ω	microseconds
μH	kΩ	nanoseconds
μH	MΩ	picoseconds

The equations relating I, V_R, V_L and time, are:

$$I = \frac{E}{R} \times (1 - e^{-Rt/L})$$

$$V_R = E \times (1 - e^{-Rt/L})$$

$$V_L = E\, e^{-Rt/L}$$

where t is the time in seconds which elapses after S is closed in position 1; $e = 2.7183$ and is the base of the natural logarithms.

In a time of L/R seconds after S is closed in position 1, the circuit conditions are: $I = 63.2\%$ of $\frac{E}{R}$, $V_R = 63.2\%$ of E, $V_L = 36.8\%$ of E and the rate of growth of current is 36.8% of E/L amperes per second (Fig. 6-7). In 2L/R seconds, the current reaches 86.5% of its final value and after 3L/R and 4L/R seconds, the corresponding percentages are 95.0% and 98.2%. When a total time of 5L/R seconds has elapsed, the transient state is assumed to have finished and the circuit has reached its final steady-state condition in which: $I = E/R$, $V_R = E$, $V_L = 0$ and the rate of growth of current is zero. The energy then stored in the magnetic field surrounding the inductor is: $0.5LI^2 = 0.5LE^2/R^2$.

If S is now switched to position 2 (Fig. 6-6), the inductor prevents any sudden change of current so that the initial conditions are: $I = E/R$, $V_R = E$, $V_L = -E$ and the rate of current decay is E/L amperes per second.

Note that the polarity of V_L is reversed compared with the conditions when S was in position 1 since the magnetic field surrounding the inductor is now collapsing. As the current decreases V_R and V_L fall together so that the rate of the current decay is less. The equations are:

$$I = \frac{E}{R} \, e^{-Rt/L}$$
$$V_R = E \, e^{-Rt/L}$$
$$V_L = -E \, e^{-Rt/L}$$

where t is the time in seconds which elapses after S is switched to position 2. Following a time of L/R seconds, the circuit conditions are:

$$I = 36.8\% \text{ of } E/R, \; V_R = 36.8\% \text{ of } E,$$
$$V_L = 36.8\% \text{ of } E.$$

The rate of current decay is 36.8% of E/L amperes per second (Fig. 6-8). In 2L/R seconds the current falls to 13.5% of E/R (and has therefore lost 86.5% of its original value). After 3L/R and 4L/R seconds, the corresponding percentage current levels are 5.0% and 1.8%. When a total time of 5L/R has elapsed, all quantities in the circuit are virtually zero.

Example 6-8

In Fig. 6-6, L = 26 mH, R = 3.3 kΩ, and E = 20 V. What is the time constant of this series combination? When S is closed in position 1, find the initial values of V_R, V_L and I and their subsequent values after time intervals of 5, 7.88, 20 and 40 microseconds. What is the energy stored in the inductor after the time interval of 40 microseconds?

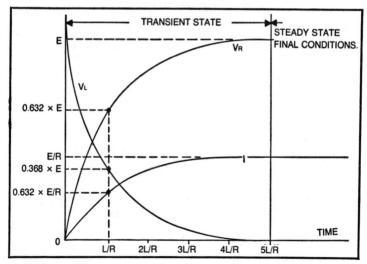

Fig. 6-7. Current growth in series LR circuit.

Solution

$$\text{Time constant} = \text{L/R} \quad \frac{26 \times 10^{-3}}{3.3 \times 10^3} \text{ seconds} = 7.88 \mu \text{ seconds.}$$

Initial values: t = 0

$I = 0$, $V_R = 0$, $V_L = 20$ V and the rate of current

growth is $E/L = 20/(26 \times 10^{-3})$

= 769 amperes per second.

t = 5 microseconds.

$$I = \frac{E}{R} (1 - e^{-Rt/L}) = \frac{20}{3.3} (1 - e^{-5/7.88}) = 2.85 \, mA.$$

$V_R = I \times R = 2.85 \, \text{mA} \times 3.3 \text{k}\Omega = 9.4 \, V.$

$V_L = E - V_L = 20 - 9.4 = 10.6V.$

t = 7.88 microseconds. This interval is equal to the time constant of the circuit.

Therefore $I = 63.2\%$ of $E/R = 0.632 \times 20/3.3 = 3.83$ mA.

$V_R = 0.632 \times 20 = 12.64 \, V.$

$V_L = 20 - 12.64 = 7.36V.$

t = 20 microseconds.

$$I = \frac{20}{3.3} (1 - e^{-20/7.88}) = 5.58 \, \text{mA.}$$

$V_R = 5.58 \, \text{mA} \times 3.3 \, \text{k}\Omega = 18.42V.$

$V_L = 20 - 18.42 = 1.58V.$

t = 40 microseconds. This interval exceeds five time constants and therefore the transient state has finished. Then: $V_R = 20$ V, $V_L = 0$ V and $I = 20$ V/3.3 kΩ = 6.06 mA. The energy stored in the inductor is:

$$\frac{1}{2} LI^2 = \frac{1}{2} \times 26 \times 10^{-3} \times (6.06 \times 10^{-3})^2$$

$$= 477.5 \times 10^{-9} \text{joules}$$

$$= 0.4775 \, microjoules.$$

Example 6-9

In Example 6-8 a time interval of 40 microseconds has elasped from the closing of the switch S in position 1. S is now switched to position 2. What are the initial values of I, V_R, V_L and the rate of current decay? What are the subsequent values of I, V_R, V_L after time intervals of 5, 7.88, 20 and 40 microseconds?

Solution

Initial values: t = 0.

$I = 6.06 \, \text{mA}, V_R = 20 \, V, V_L = -20 \, V.$

$$t = 5 \text{ microseconds}.$$

$$I = \frac{E}{R} \; e^{-Rt/L} = 6.06 \times e^{-5/7.88} = 3.21 \text{ mA}.$$

$$V_R = I \times R = 3.21 \text{ mA} \times 3.3 \text{ k}\Omega = 10.6 \text{ V}.$$
$$V_L = -V_R = -10.6 \text{ V}.$$

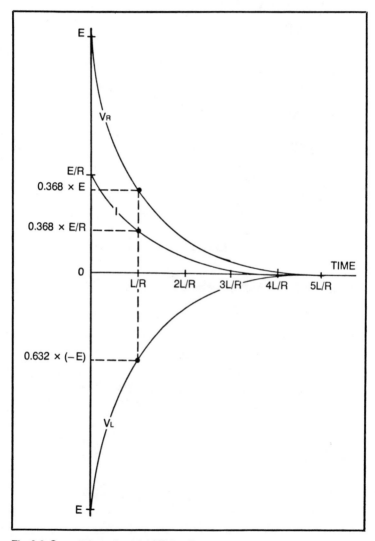

Fig. 6-8. Current decay in series LR circuit.

$t = 7.88$ *microseconds*.

    $I = 36.8\%$ of $E/R = 0.368 \times 6.06$ mA = 2.23 mA.

$V_R = 2.23$ mA $\times 3.3$ k$\Omega = 7.36$ V.

$V_L = -7.36$ V.

$t = 20$ *microseconds*.

    $I = 6.06 \times e^{-20/7.88} = 0.48$ mA.

    $V_R = 0.48$ mA $\times 3.3$ k$\Omega = 1.58$ V.

    $V_L = -1.58$ V.

$t = 40$ *microseconds*. $I = 0$ A, $V_R = V_L = 0$ V.

CHAPTER SUMMARY

☐ Induced EMF $= -L\dfrac{\Delta I}{\Delta T} = -N\dfrac{\Delta\phi}{\Delta T}$ volts.

☐ Inductance, $L = N\dfrac{\Delta\phi}{\Delta I}$ henrys.

☐ Inductance of coil, $L = \dfrac{\mu_0 \mu r\, N^2 A}{\iota}$ henrys.

☐ Energy stored in magnetic field $= \frac{1}{2}LI^2$ joules.

☐ Time constant of LR circuit $= \dfrac{L}{R}$ seconds.

☐ Time required to reach steady-state conditions $= \dfrac{5L}{R}$

Electrostatics and Capacitance

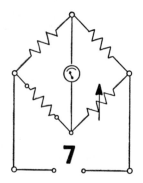

7

In 1785 Coulomb discovered through experiments that the force between two charged bodies was directly proportional to the product of their charges and inversely proportional to the square of their separation.

COULOMB'S LAW

In equation form:

$$F = \frac{Q_1 Q_2}{4 \pi \epsilon_0 d^t} = k \frac{Q_1 Q_2}{d^2} \quad \text{newtons}$$

where: Q, Q_2 = charges measured in coulombs

and: d = distance between charges in meters

$k = 1/(4\pi \epsilon_0) = 1/(4\pi \times 8.85 \times 10^{-12})$
$= 8.99 \times 10^9$ SI units.

The meaning of ϵ_0, the permittivity of free space, is explained later in this chapter. In the cgs system the unit of charge is the *statcoulomb*. One coulomb = 3×10^9 statcoulombs. Between the charges exists an electric field whose total number of lines is referred to as the electric flux. In the SI system a single line of the electric flux is assumed to leave a positive charge of 1 coulomb and terminate on a negative charge of 1 coulomb. The number of lines is therefore equivalent to the number of coulombs and no special unit is required for the electric flux whose symbol is the Greek letter, ψ (psi).

Figure 7-1 shows two rectangular surface plates which are separated by a vacuum insulator of thickness d meters. For each plate the area of one side is A meters. The plates have been charged to a potential difference of V_c volts so that one plate carries a positive charge of Q coulombs and the other plate has an equal negative charge. The electric flux therefore consists of Q lines and the electric flux density, D, in MKS units is given by:

$$D = \frac{Q}{A}$$

coulombs per square meters. In the cgs system electric flux density is measured in statcoulombs per square centimeter. One coulomb per square meter = $12\,\pi \times 10^5$ statcoulombs per square centimeter.

In the region between the plates the electric field intensity, ϵ, is equal to the voltage gradient so that:

$$\epsilon = \frac{V_c}{d}\ \text{volts per meter.}$$

ELECTRIC FIELD INTENSITY

If a charge of 1 coulomb is positioned between the plates, an electric field intensity of one volt per meter will cause a force, F, of one newton to be exerted on the charge. Therefore:

$$\epsilon = \frac{F}{Q}$$

and the field intensity may also be measured in newtons per coulomb.

The cgs unit of electric field intensity is the statvolt per centimeter, where

$$1\ \text{volt per meter} = \frac{1}{3} \times 10^{-4}\ \text{statvolts per centimeter.}$$

The electric field intensity can be compared with the magnetic field intensity, H, which is measured in amperes per meter.

In electromagnetism the ratio of the magnetic flux density B, in a vacuum to the magnetic field intensity, H, is the *permeability* of free space, $\mu_o = 4\,\pi \times 10^{-7}$ mks units. Likewise in electrostatics the ratio of the electric flux density in a vacuum to the electric field intensity is called the *permittivity* of free space whose letter symbol is ϵ_o. In equation form

$$\frac{D}{\epsilon} = \epsilon_o$$

The value of ϵ_o is 8.85×10^{-12} mks units which could be expressed in coulombs per volt meter. In the cgs system $\epsilon_o = 1$.

With $\mu_o = 4\,\pi \times 10^{-7}$ and $\epsilon_o = 8.85 \times 10^{-12}$, it may be shown that:

$$\frac{1}{\sqrt{\mu_o \epsilon_o}} = 3 \times 10^8\ \text{meters per second, which is the}$$

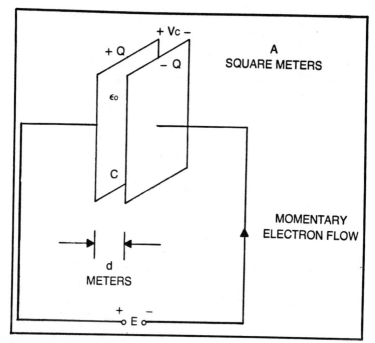

Fig. 7-1. The principle of the capacitor.

velocity of all electromagnetic waves. One example of such a wave is light. In addition,

$$\sqrt{\frac{\mu_{o}}{\epsilon_{o}}} = 120\,\pi = 377\,\Omega$$

This is the value of the free space impedance, η_{o}. Clerk Maxwell discovered these relationships in 1865 and he was then able to predict the existence of radio waves some twenty years before the experiments of Heinrich Hertz.

CAPACITANCE AND THE CAPACITOR

Capacitance is that property of an electrical circuit which opposes any sudden change of *voltage*. This compares with inductance which opposes any sudden change of *current*, and resistance which merely limits the flow of current.

The property of capacitance is possessed by a capacitor which consists of two conducting surfaces, (for example aluminum, tin foil or copper) separated by an insulator (air, mica or paper) which

111

is referred to as the dielectric. Figure 7-1 therefore represents a capacitor with a vacuum as the dielectric. When a DC voltage, E, is applied across the capacitor there is a momentary electron flow which ceases when the voltage between the plates exactly balances the applied voltage. The charge then stored depends on this voltage and some constant of the capacitor called its capacitance, C, measured in Farads (after Michael Faraday). In equation form:

$$V_c = - \frac{Q}{C}.$$

The negative sign indicates that V_c balances the appled voltage, E. By Kirchhoff's Voltage Law: $E + V_c = 0$. Therefore:

$$E = \frac{Q}{C}.$$

The capacitance is one farad if when the voltage between the capacitor plates is one volt, the charge stored is one coulomb. However, the farad is too large for practical purposes so that capacitors are normally measured in either microfarads or picofarads.

$$\epsilon_0 = \frac{D}{\epsilon} = \frac{Q/A}{V_c/d} = \frac{Q}{V_c} \times \frac{d}{A} = C \times \frac{d}{A}$$

ϵ_0 can therefore be measured in farads per meter as well as coulombs per volt meter. Then

$$C = \frac{\epsilon_0 A}{d} \text{ farads.}$$

Note that the capacitance is directly proportional to the area of the plates but is inversely proportional to their separation.

Since $Q = C \times V_c$, it is impossible to change the voltage between the capacitor plates without altering the value in coulombs of the charge stored. This required that a current in amperes or coulombs per second shall either charge or discharge the capacitor over a period of time. In other words, a capacitor can neither charge or discharge instantaneously so that capacitance is that electrical property which opposes any sudden change of voltage.

RELATIVE PERMITTIVITY

If the space between the capacitor plates is filled by a particular insulating material rather than a vacuum, the capacitance

Table 7-1. Dielectric Constants.

Material	Relative permittivity, μ_r
Air	1.0006
Alumina	4.5 - 8.4
Beeswax	2.66
Epoxy cast resin	3.62
Mica	5.4
Neoprene	6.60
Nylon	3.5
Porcelain	6.0 - 8.0
Quartz (fused)	3.75 - 4.1
Silica glass	3.8
Steatite	5.5 - 7.5
Teflon	2.0
Titanium dioxide	14 - 110

is increased. The ratio of a capacitor's capacitance with a certain material as the dielectric to the capacitance with a vacuum dielectric is called the material's relative permittivity or dielectric constant, ϵ_r.

Table 7-1 shows the values of the dielectric constant for some commonly used insulating materials.

If the space between the capacitor plates is filled by a dielectric with a relative permittivity, ϵ_r,

$$C = \frac{\epsilon_o \, e_r A}{d} \text{ farads.}$$

and:

$$\frac{\text{Electric flux density, D}}{\text{Electric field intensity, } \epsilon} = \epsilon_o \epsilon_r = \epsilon$$

ϵ is the absolute permittivity.

$$\frac{\text{Magnetic flux density, B}}{\text{Magnetic field intensity, H}} = \mu_o \mu_r = \mu$$

where μ is the absolute permittivity. Compare this equation with the one above.

Example 7-1

A capacitor consists of two sheets of tin foil, each side of which has an area of 1500 cm^2. The sheets are separated by a 0.06 mm thickness of paper dielectric whose relative permittivity is 2.25. What is the value of the capacitance?

Solution

Capacitance, $C = \dfrac{\epsilon_o \epsilon_r A}{d} = \dfrac{8.85 \times 10^{-12} \times 2.25 \times 1500 \times 10^{-4}}{0.06 \times 10^{-3}}$

$$= \dfrac{8.85 \times 2.25 \times 1.5 \times 10^{-2}}{6} \ \mu F$$

$$= 0.0498 \ \mu F.$$

Example 7-2

The capacitor in Example 7-1 is charged to 80 V. Calculate the values of the (a) charge stored, (b) electric flux density and (c) electric field intensity.

Solution

(a) Charge stored, $Q = C \times V_c = 0.0498 \times 10^{-6} \times 80$ C
$$= 3.98 \text{ microcoulombs.}$$

(b) Electric flux density, $D = \dfrac{Q}{A} = \dfrac{3.98}{1500 \times 10^{-4}}$

$$= 26.6 \text{ microcoulomb}$$
per square meter.

(c) Electric field intensity, $= \dfrac{V_c}{d} = \dfrac{80}{0.06 \times 10^{-3}}$

$$= 1.33 \times 10^6 \text{ volts per meter.}$$

Check: $\dfrac{D}{\epsilon} = \dfrac{26.6 \times 10^{-6}}{1.333 \times 10^6} = 19.9 \times 10^{-12} \text{ farads per meter}$

Absolute permittivity, $\epsilon = \epsilon_o \epsilon_r = 8.85 \times 10^{-12} \times 2.25$

$$= 19.9 \times 10^{-12} \text{ farads per meter.}$$

ENERGY STORED IN THE ELECTRIC FIELD BETWEEN THE CAPACITOR PLATES

Let a capacitor of capacitance C farads be charged by a constant current I amperes over a period of t seconds (Fig. 7-2). The final charge, Q is I × t coulombs and the corresponding voltage, V_c, between the capacitor plates is Q/C. Since the charging current is constant, the average voltage is $kV_c/2$.

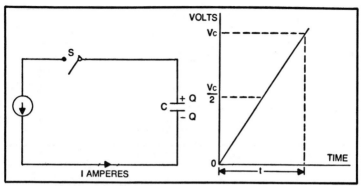

Fig. 7-2. Energy stored in a capacitor.

Energy supplied during charging = power × time

$$= I \times \frac{V_C}{2} \times t$$

$$= \frac{1}{2} V_C Q = \frac{1}{2} \frac{Q^2}{C}$$

$$= \frac{1}{2} CV_C^2 \text{ joules.}$$

CAPACITORS IN SERIES

Referring to Fig. 7-3 the capacitors C_1, C_2, C_3 are in series and are charged by the source voltage, E_T. During the charging of the capacitors the current is the same throughout the circuit and therefore at the end of the charging period, each capacitor must carry the same charge, Q. Then:

$$V_1 = \frac{Q}{C_1}, V_2 = \frac{Q}{C_2} \text{ and } V_3 = \frac{Q}{C_3}$$

Also: $E_T = V_1 + V_2 + V_3$

$$= Q \times \left(\frac{1}{C_1} + \frac{1}{C_2} + \frac{1}{C_3} \right).$$

If C_T is the total equivalent capacitance which will store the same charge Q:

$$E_T = \frac{Q}{C_T} = Q \times \frac{1}{C_T} = V_1 \times \frac{C_1}{C_T}$$

Therefore: $V_1 = E_T \times \frac{C_T}{C_1}, V_2 = E_T \times \frac{C_T}{C_1}, V_3 = E_T \times \frac{C_T}{C_3}$

115

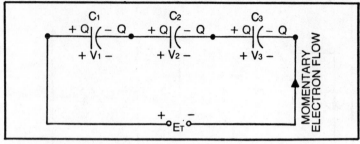

Fig. 7-3. Capacitors in series.

and: $\dfrac{1}{C_T} = \dfrac{1}{C_1} + \dfrac{1}{C_2} + \dfrac{1}{C_3}$

For N capacitors in series:

$$C_T = \cfrac{1}{\dfrac{1}{C_1} + \dfrac{1}{C_2} + \dfrac{1}{C_3} + \cdots + \dfrac{1}{C_N}}$$

If the N series capacitors all have the same value, C: $C_T = \dfrac{C}{N}$

For two capacitors in series:

$$C_T = \dfrac{C_1 C_2}{C_1 + C_2} \ , V_1 = E_T \times \dfrac{C_2}{C_1 + C_2} \ , V_2 = E_T \times \dfrac{C_1}{C_1 + C_2}.$$

The formulas for capacitors in series and resistors in parallel are therefore comparable. Connecting capacitors in series reduces the capacitance, and the total capacitance is less than the value of the smallest series capacitance. Basically this is because the series arrangement effectively increases the distance between the end plates connected to the source voltage and the capacitance is inversely proportional to this distance. Since a capacitor has a DC working voltage rating, the series arrangement may be used to distribute the source voltage between the capacitors so that the voltage across an individual capacitor does not exceed its rating.

CAPACITORS IN PARALLEL

In Fig. 7-4 the capacitors C_1, C_2, C_3 are in parallel and each is charged by the source voltage E_T. Therefore: $Q_1 = C_1 E_T$, $Q_2 = C_2 E_T$, $Q_3 = C_3 E_T$

The total charge, Q_T is the sum of the individual charges stored in the capacitors. In equation form: $Q_T = Q_1 + Q_5 + Q_3 = E_T (C_1 + C_2 + C_3)$.

If C_T is the total capacitance: $Q_T = E_T C_T$.

Therefore: $C_T = C_1 + C_2 + C_3$.

For N capacitors in parallel: $C_t = C_1 + C_2 + C_3 + ----- + C_N$.

If N parallel capacitors all have the same value, C_1 then $C_T = NC$.

The total equivalent capacitance of capacitors in parallel is the sum of their individual capacitances. The parallel arrangement increases the effective surface area and therefore the capacitance, which is directly proportional to A.

If a capacitor is made up of N parallel plates with alternate plates connected as in Fig. 7-5, the total capacitance is:

$$C_T = \frac{\epsilon_o \epsilon_r (N-1)A}{d}$$

where ϵ_o, ϵ_r, A and d defined as before.

Example 7-3

In Fig. 7-3, $C_1 = 10\,\mu F$, $C_2 = 16\,\mu F$, $C_3 = 8\,\mu F$ and $E_T = 56\,V$. Find the values of C_T, V_1, V_2, V_3. What is the amount of charge and energy stored in each capacitor?

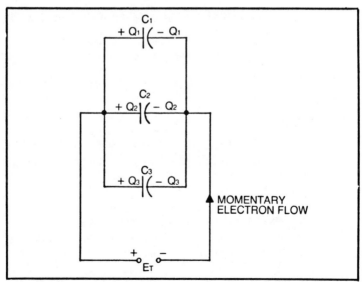

Fig. 7-4. Capacitors in parallel.

Solution

Total capacitance, $C_T = \dfrac{1}{\dfrac{1}{10} + \dfrac{1}{16} + \dfrac{1}{8}}$

$= 3.478\ \mu F.$

Charge stored in eash capacitor, $Q = 3.478 \times 10^{-6}\,F \times 56V$

$= 194.8\ \mu C.$

Then:

$V_1 = \dfrac{194.8\ \mu C}{10\ \mu F} = 19.48\,V.$

$V_2 = \dfrac{194.8\ \mu C}{16\ \mu F} = 12.17V.$

$V_3 = \dfrac{194.8\ \mu C}{8\ \mu F} = 24.35V.$

Check: $E = V_1 + V_2 + V_3 = 19.48 + 12.17 + 24.35 = 56\,V.$

Energy stored in the $10\ \mu F$ capacitor $= \dfrac{1}{2} \times 194.8\ \mu C \times 19.48\,V$

$= 1897\ \mu\,joules.$

Energy stored in the $16\ \mu F$ capacitor $= \dfrac{1}{2} \times 194.8\ \mu C \times 12.17\,V$

$= 1185\ \mu\,joules.$

Energy stored in the $8\ \mu F$ capacitor $= \dfrac{1}{2} \times 194.8\ \mu C \times 24.35\,V$

$= 2372\ \mu\,joules.$

Total energy stored $= 1897 + 1185 + 2372 = 5454\ \mu\,joules.$

Check: Total energy stored $= \dfrac{1}{2} \times 194.8\ \mu C \times 56\,V = 5454$ $\mu\,joules.$

Example 7-4

In Fig. 7-4, $C_1 = 10\ \mu F$, $C_2 = 16\ \mu F$, $C_3 = 8\ \mu F$ and $E_T = 56\,V$. Find the values of C_T, Q_T, Q_1, Q_2, Q_3. What is the energy stored in each capacitor?

Solution

Total capacitance, $C_T = 10 + 16 + 8 = 34\ \mu F.$
Total charge stored, $Q_T = 34\ \mu F \times 56\,V = 1904\ \mu C.$
Then: $Q_1 = 10\ \mu F \times 56\,V = 560\ \mu C.$
$Q_2 = 16\ \mu F \times 56\,V = 896\ \mu C.$
$Q_3 = 8\ \mu F \times 56\,V = 448\ \mu C.$

Check: Total charge stored, $Q_T = Q_1 + Q_2 + Q_3$
$$= 560 + 896 + 448 = 1904$$
μC.

Energies stored in C_1, C_2, C_3 are respectively:
$\frac{1}{2} \times 560 \times 56$, $\frac{1}{2} \times 896 \times 56$, $\frac{1}{2} \times 448 \times 56$ or *15680, 25088, 12544* μjoules.

Total energy stored is $15680 + 25088 + 12544 = 53312$ μjoules.

Check: Total energy stored is $\frac{1}{2} \times Q_T \times E_T = \frac{1}{2} \times 1904 \,\mu\text{C} \times 56 \text{ V}$
$$= 53312 \ \mu\text{joules.}$$

Example 7-5

A 40 μF capacitor is charged from a 200 V source. Another 20 μF capacitor is charged from a 100V source. Immediately after the capacitors are disconnected, they are correctly paralleled. What is (a) the voltage across the combination and (b) the total energy before and after the parallel connection?

Solution

(a) Charge stored in the 40 μF capacitor = 200 V $\times$ 40 μF
$$= 8000 \ \mu\text{C.}$$
Charge stored in the 20 μF capacitor = 100 V $\times$ 20 μF
$$= 2000 \ \mu\text{C.}$$
Total charge = $8000 + 2000 = 10000 \ \mu$C.
Total capacitance of the parallel combination
$$= 40 \ \mu\text{F} + 20 \ \mu\text{F}$$
$$= 60 \ \mu\text{F.}$$

Voltage across the parallel combination $= \dfrac{10000}{60} = 166.7 \text{ V.}$

(b) Energy stored in the 40 μF capacitor
$$= \frac{1}{2} \times 8000 \ \mu\text{C} \times 200 \text{ V}$$
$$= 0.8 \text{ joules}$$
Energy stored in the 20 μF capacitors
$$= \frac{1}{2} \times 2000 \ \mu\text{C} \times 100 \text{ V}$$
$$= 0.1 \text{ joule.}$$
Total energy stored before the parallel connection = $0.8 + 0.1 = 0.9$ joule.

Total energy stored after the parallel connection $= \frac{1}{2} \times 10000$ C $\times$ 166.7 V = 0.833 joule.

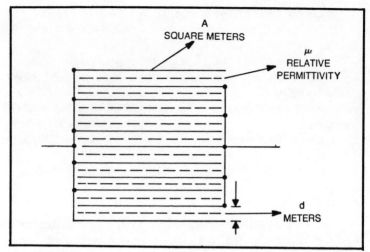

Fig. 7-5. Multi-plate capacitor.

Note that energy has been lost as the result of the parallel connection. Neglecting the resistance of any connecting wires, the lost energy appears in the form of the spark which appears when the parallel connection is made.

Example 7-6

A capacitor is made with nine conducting surfaces as shown is Fig. 7-5. The area of one side of one surface is 150 cm^2 and the surfaces are separated by sheets of mica with a thickness of 0.4mm and a relative permittivity of 5. What is the value of the capacitance?

Solution

$N = 9, A = 150 \times 10^{-4} m^2$, $d = 0.4 \times 10^{-3} m$, $\epsilon_r = 5$.
Then:

$$C = \frac{\epsilon_o \epsilon_r^{(N-1)A}}{d} = \frac{8.85 \times 10^{-12} \times 5 \times 8 \times 150 \times 10^{-4}}{0.4 \times 10^{-3}}$$

$$= 8.85 \times 10 \times 150 \; \rho F$$
$$= 0.0\bar{1}33 \; \mu F.$$

TIME CONSTANT OF A CAPACITOR AND A RESISTOR IN SERIES

In the circuit of Fig. 7-6, the capacitor, C, is uncharged. The circuit conditions immediately after the switch S is closed in

position 1, are
$$Q = 0, V_c = 0, V_R = E$$
and the charging current,
$$I_c = \frac{E}{R} \quad \text{(maximum value)}$$

Q is the charge, measured in coulombs, which must initially be zero since a capacitor can neither charge nor discharge instantaneously. Since no voltage can appear across the capacitor until time has elapsed after the switch is closed in position 1, the capacitor must initially behave as a *short* circuit.

The time constant of the circuit is CR seconds, where C is measured in farads and R in ohms. After a time interval of CR seconds from the instant S is closed in position 1, the circuit conditions as shown in Fig. 7-7 are:
$$Q = 63.2\% \text{ of CE coulombs}$$
$$V_c = 63.2\% \text{ of E}$$
$$V_R = 36.8\% \text{ of E}$$
$$I_c = 36.8\% \text{ of } \frac{E}{R}$$

The order of the time constant value may be found from Table 7-2. Also, see Table 7-3.

Following an interval of 2CR seconds from the instant S is closed in position 1, the capacitor has acquired 86.5% of its final charge, CE coulombs, (and the voltage across the capacitor is 0.865 E). After 3CR and 4CR seconds, the charge percentages are respectively 95.0% and 98.2%. After 5CR seconds, the circuit is assumed to have reached its steady state conditions which are Q =

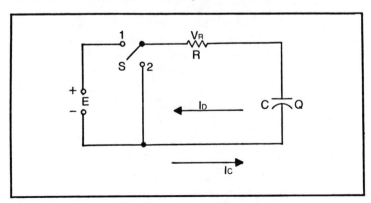

Fig. 7-6. Charge and discharge of a capacitor through a resistor.

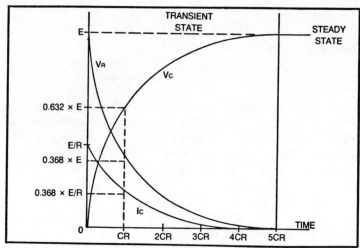

Fig. 7-7. Charging of a capacitor through a resistor.

CE, $V_C = E$, $V_R = 0$, $I_C = 0$. The fully charged capacitor then behaves as an *open* circuit. The following equations refer to the charging of the capacitor during the transient state.

$$Q = CE(1 - e^{-t/CR})$$
$$V_C = E(1 - e^{-t/CR})$$
$$I_C = \frac{E}{R} e^{-t/CR}$$
$$V_R = E e^{-t/CR}$$

where e is $= 2.7183$ and is the base of the natural logarithms.

If S is now switched to position 2, the initial conditions are:
$$Q = CE, V_C = E, V_R = -E, I_D = -E/R$$

The negative signs for the values of V_R and I_D indicate that the polarity of the voltage drop across the resistor, V_R, has reversed since the discharge current, I_D, is in the opposite direction to the previous charging current, I_{oc}; I_{gd} and V_R are therefore shown below the time axis in Fig. 7-8. After an interval equal to one time constant (CR seconds), the conditions are:

$$Q = 36.8\% \text{ of } CE$$
$$V_C = 36.8\% \text{ of } E$$
$$V_R = 36.8\% \text{ of } (-E)$$
$$I_D = 36.8\% \text{ of } (-E/R)$$

Following an interval of 2CR from the instant S is switched to position 2, the capacitor's charge has fallen to 13.5% of CE

Table 7-2. CR Time Constants.

C	R	Time Constant CR
farads	ohms	seconds
μF	ohms	microseconds
μF	kΩ	milliseconds
μF	MΩ	seconds
pF	ohms	picoseconds
pF	kΩ	nanoseconds
pF	MΩ	microseconds

coulombs (the capacitor has therefore lost 86.5% of its initial charge). After 3CR seconds, the percentages for the capacitor's remaining charge are respectively 5.0% and 1.8%. The equations for the transient discharge state are:

$$Q = CE \, e^{-t/CR}$$

$$V_C = E \, e^{-t/CR},$$

$$I_D = \frac{-E}{R} \, e^{-t/CR}$$

$$V_R = -E \, e^{-t/CR}$$

Table 7-3. Converting SI/cgs Units.

QUANTITY	SI UNITS	cgs ELECTROSTATIC UNITS
Charge, Q	1 coulomb	3×10^9 statcoulombs
EMF, E, and P.D. V	1 volt	1/300 statvolt
Current, I	1 ampere	3×10^9 statamperes
Resistance, R	1 ohm	1.11×10^{-12} statohms
Capacitance, C,	1 farad	9×10^{11} statfards
Electric field intensity, ϵ	1 volt/meter 1 newton/coulomb	$\frac{1}{3} \times 10^{-4}$ statvolt per centimeter
Electric flux density, D,	1 coulomb per square meter	$12\pi \times 10^5$ statcoulombs per square centimeter
Permittivity of free space, ϵ_0	8.85×10^{-12} farads per meter	1

In a total time of 5CR seconds, the capacitor is assumed to have fully discharged and all circuit quantities are zero.

Example 7-7

In Fig. 7-6, C = 50 pFs, R = 1 MΩ and E = 80 V. What is the time constant of the circuit? If S is closed in position 1, what are the values of Q, V_C, V_R, I_C after time intervals of 0, 30, 50, 90, 180 and 300 μseconds. What is the energy stored in the capacitor after 250 μseconds?

Solution

Time constant = C × R = *50* μseconds.
Initial conditions (t = 0).

$$V_C = 0 \text{ V, } V_R = 80 \text{ V, } Q = 0 \text{ C, } I_C = \frac{80 \text{ V}}{1 \text{ M}\Omega} = 80 \text{ }\mu\text{A.}$$

Time = 30 μ seconds.
$$V_C = E(1 - e^{-t/CR}) = 80(1 - e^{-30/50}) = 36.1 \text{ V.}$$
$$V_R = E - V_C = 80 - 36.1 = 43.9 \text{ V.}$$
$$Q = C \times V_C = 50 \text{ pFs} \times 36.1 \text{ V} = 1805 \text{ pC.}$$

$$I_C = \frac{V_R}{R} = \frac{43.9 \text{ V}}{1 \text{ M}\Omega} = 43.9 \text{ }\mu\text{A.}$$

Time = 50 μ seconds (one time constant).
$$V_c = 80 \times 0.632 = 50.6 \text{ V.}$$
$$V_R = 80 - 50.6 = 29.4 \text{ V.}$$
$$Q = 50 \text{ pFs} \times 50.6 \text{ V} = 2530 \text{ pC.}$$

$$I_C = \frac{29.4 \text{ V}}{1 \text{ M}\Omega} = 29.4 \text{ }\mu\text{A.}$$

Time = 90 μ seconds;.
$$V_c = 80(1 - e^{-90/50}) = 66.8 \text{ V.}$$
$$V_R = 13.2 \text{ V.}$$
$$Q = 50 \text{ pF} \times 66.8 \text{ V} = 3340 \text{ pC.}$$

$$I_C = \frac{13.2 \text{ V}}{1 \text{ M}\Omega} = 13.2 \text{ }\mu\text{A.}$$

Time = 180 μ seconds.
$$V_C = 80(1 - e^{-180/50}) = 77.8 \text{ V.}$$
$$V_R = 2.2 \text{ V.}$$
$$Q = 50 \text{ pF} \times 77.8 \text{ V} = 3890 \text{ pC.}$$

$$I_C = \frac{2.2 \text{ V}}{1 \text{ M}\Omega} = 2.1 \text{ mA.}$$

Time = 300 μ seconds (six time constants).

The circuit is now in its steady state condition.

$V_C = 80\,V.$
$V_R = 0\,V.$
$Q = 50\,pF \times 80\,V = 4000\,pC.$
$I_C = 0\,A.$

Energy stored in the capacitor after 250 μseconds (five time constants) $= \frac{1}{2}C\,E^2 = \frac{1}{2} \times 50 \times 10^{-12} \times 80^2$ joule.

$$= 1.6 \times 10^{-7} \text{joule.}$$

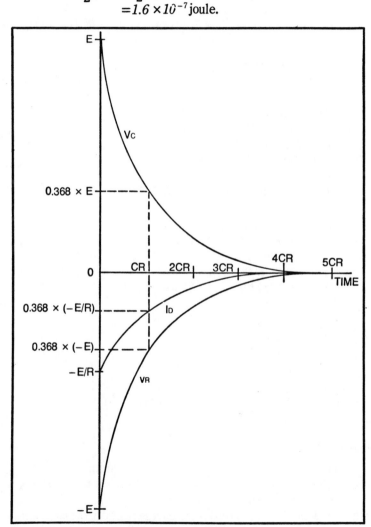

Fig. 7-8. Discharge of a capacitor through a resistor.

Example 7-8

In Fig. 7-6, S is closed in position 1 and the circiit has reached its steady state condition. S is now switched to position 2. From that instant what are the values of Q, V_c, V_R, I_D after time intervals of 0, 30, 50, 90, 180 and 300 μseconds?

Solution

Initial conditions (t = 0).

$$V_{gc} = 80\,V, V_R = 80\,V, Q = 50\,pFs \times 80\,V = 4000\,pC \text{ and } I_D$$
$$= -80\,V/1\,M\Omega = -80\,\mu A.$$

Time = 30 μ seconds.
$V_C = E\,e^{-t/CR} = 80 \times e^{-30/50} = 43.9\,V.$
$V_R = -43.9\,V.$
$Q = 50\,pFs \times 43.9\,V = 2195\,\mu C.$
$I_D = -43.9\,V/1\,M\Omega = -43.9\,\mu A.$

Time = 50 μ seconds (one time constant).
$V_C = 80 \times 0.368 = 29.4\,V.$
$V_R = -29.4\,V.$
$Q = 50\,pFs \times 29.4\,V = 1470\,pC.$
$I_D = -29.4\,V/1\,M\Omega = -29.4\,\mu A.$

Time = 90 μ seconds
$V_C = 80 \times e^{-90/50} = 13.2\,V.$
$V_R = 13.2\,V.$
$Q = 50\,pFs \times 13.2\,V = 661\,pC.$
$I_D = 13.2\,V/1\,M\Omega = -13.2\,\mu A.$

Time = 180 μ seconds.
$V_C = 80 \times e^{-180/50} = 2.2\,V.$
$V_R = -2.2\,V.$
$Q = 50\,pFs \times 2.2\,V = 110\,pC.$
$I_D = 2.2\,V/1\,M\Omega = -2.2\,\mu A.$

Time = 300 μ seconds (six time constants).
The capacitor is now completely discharged. $Q = 0\,C$, $V_C = V_R = 0\,V$, $I_D = 0\,A.$

Example 7-9

In Fig. 7-9 the capacitor is initially uncharged and the switch S is then closed. What are the values of I_1, I_2, I_3 and the potential at X at the start and finish of the transient state? What is the time constant of the circuit and what is the energy stored in the capacitor at the end of the transient state?

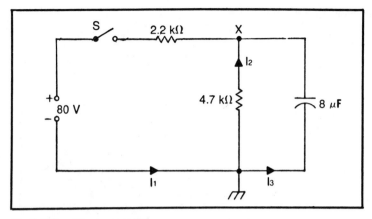

Fig. 7-9. Circuit for example 7-9.

Solution

Since the capacitor cannot charge instantaneously the initial potential at X is *zero* volts and $I_2 = 0\,A$. Therefore $I_1 = I_3 = 80\,V/2.2$ kV = *36.4 mA*. At the end of the transient state the capacitor is fully charged and $I_3 = 0\,A$. Then $I_1 = I_2 = 80\ V/(2.2\ k\Omega + 4.7\ k\Omega) =$ *11.6 mA* and the potential at X is 11.6 mA × 4.7 kΩ =$\pm54.5\ V$. This is also the voltage to which the capacitor is charged so that the final energy stored is $\frac{1}{2} \times 8\ \mu F \times (54.5\ V)^2 =$ *11890 µjoules*.

With respect to the capacitor the resistors are in parallel (this may be shown by the use of Thévenin's Theorem in chapter eight.

The circuits time constant is $\dfrac{2.2 \times 4.7\ k\Omega}{2.2 + 4.7} \times 8\ \mu F =$ *12 milliseconds*, rounded off.

Example 7-10

In Fig. 7-10 the capacitors are initially uncharged and the switch S is then closed. What are the values of I and the potentials at X, Y, Z at the start and at the finish of the transient state? What is the value of the circuit's time constant?

Solution

Since the capacitors cannot charge instantaneously, the supply voltage of 110 V must initially be divided between the 3.3 kΩ and 4.7 kΩ resistors. Therefore I is 110 V/(3.3 kΩ + 4.7 kΩ) = 13.75 mA and the potentials at X, Y, Z are all + 13.75 mA × 4.7 kΩ = 64.6 V.

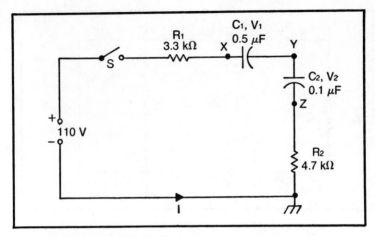

Fig. 7-10. Circuit for example 7-10.

At the end of the transient state the capacitors are fully charged and therefore I must be zero. The 110 V must now be divided between the capacitors so that:

$$V_2 = E_T \times \frac{C_1}{C_1 + C_2} = 110\,V \times \frac{0.5\,\mu F}{0.5\,\mu F + 0.1\,\mu F} = 91.7\,V.$$

The potentials at X, Y, and Z are respectively: $+110\,V$, $+91.7\,V$, $0\,V$.

The total capacitance is $\frac{0.5 \times 0.1}{0.5 + 0.1} = 0.0833\,\mu F$ so that the circuit's time constant is $(3.3 + 4.7)\,k\Omega \times 0.0833\,\mu F = 0.66$ *millisecond*.

CHAPTER SUMMARY

☐ Q (coulombs) = C (Farads) × V_c (volts)

☐ Electric Field Intensity, $\epsilon = \dfrac{V_c}{d}$ volts per meter or newtons per coulomb.

☐ Electric Flux Density, $D = \dfrac{Q}{A}$ coulombs per square centimeter.

☐ Absolute permittivity, $e = \dfrac{D}{\epsilon} = \epsilon_o \epsilon_r$ farads per meter.

☐ Relative permittivity or dielectric constant of an insulating material. ϵ_r

$$= \dfrac{\text{capacitance with insulating materials as the dielectric}}{\text{capacitance with a vacuum dielectric}}$$

☐ Capacitance, $C = \dfrac{\epsilon_o \epsilon_A}{d} = \dfrac{\epsilon A}{d}$ farads.

☐ Energy stored in a capacitor $= \dfrac{1}{2} C V_c^2 = \dfrac{1}{2} Q V_c = \dfrac{1}{2} \dfrac{Q^2}{C}$

☐ Capacitors in series: $\dfrac{1}{C_T} = \dfrac{1}{C_1} + \dfrac{1}{C_2} + \dfrac{1}{C_3} + \cdots \dfrac{1}{C_N}$

$$V_1 = E_T \times \dfrac{C_T}{C_1} , V_2 = E_T \times \dfrac{C_T}{C_2} , \text{etc}$$

☐ Two capacitors in series:

$$C_T = \dfrac{C_1 C_2}{C_1 + C_2} , V_1 = E_T \times \dfrac{C_2}{C_1 + C_2} , V_2 = E_T \times \dfrac{C_1}{C_1 + C_2}$$

☐ Capacitors in parallel:

$$C_T = C_1 + C_2 + C_3 \cdots + C_N$$
$$Q_T = Q_1 + Q_2 + Q_3 \cdots + Q_N$$

☐ Multiplate Capacitor:

$$C = \dfrac{\epsilon_o \epsilon_r (N-1) A}{d}$$

☐ Charge and Discharge of Capacitor through Resistor:
Time constant in seconds = C (farads) × R (ohms) Total time of transient state = 5 CR seconds.

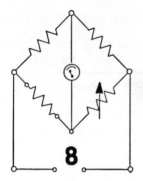

Network Theorems

8

An electrical resistor network may be defined as any electrical network containing resistors and sources. A single network may consist of a single closed circuit (or mesh) while more complex networks could contain a number of meshes which are interdependent.

The current through, and the voltage across, any resistor of a network may be found by applying Ohm's and Kirchhoff's Laws but, in the case of a complex network, the process is lengthy and tedious because of the need to solve a large number of simultaneous equations. A series of network theorems have therefore been formulated to simplify the calculations.

Some of the theorems in this chapter have universal application while others are restricted to circuits containing linear resistances. A *linear resistance* is defined as any resistance which obeys Ohm's Law so that the voltage across the resistance is directly proportional to its current. Resistors fall into this category while semiconductor devices and tubes do not.

In this chapter the theorems are only used with dc circuits although they are equally valid for ac circuits. However before the theorems can be applied to general ac networks, it would be necessary to have a background of complex algebra (Chapter 13).

The following theorems are stated without proof, but examples are used to illustrate their application.

SUPERPOSITION THEOREM

If a network of *linear* resistances contains more than one source, the current flowing at any point is the algebraic sum of the currents that would flow at that point if each source was considered

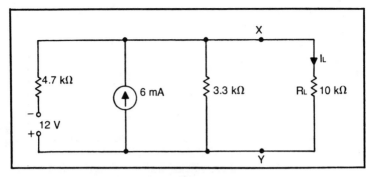

Fig. 8-1. Circuit illustrating various network theorems.

separately while at the same time all other sources were replaced by their equivalent internal resistances. This last step would be achieved by short-circuiting all sources of constant voltage and open-circuiting all sources of constant current.

Example 8-1

In the circuit of Fig. 8-1, calculate the value of the load current, I_L.

Solution

The problem may be solved in a number of ways.

Method 1: Millman's Theorem: (see chapter 4)

Convert the 12 V source into the equivalent current generator (Fig. 8-2). The constant current is $\dfrac{12\,V}{4.7k\Omega} = 2.55\,mA$ so that the total generator current is $6\,mA + 2.55\,mA = 8.55\,mA$. The total equivalent internal resistance is

$$\frac{4.7 \times 3.3}{4.7 + 3.3} = 1.94k\Omega.$$

Then by the current division rule: Load current, $I_L = 8.55\,mA \times \dfrac{1.94k\Omega}{10k\Omega + 1.94\,k\Omega} = 1.39\,mA.$

Method 2: Kirchhoff's Voltage and Current Laws.

Convert the 6 mA constant current generator into its constant voltage equivalent. The constant voltage is $6\,mA \times 3.3\,k\Omega = 19.8$

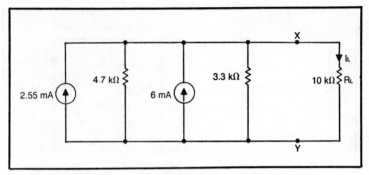

Fig. 8-2. Solution by Millman's theorem.

V which is in series with the 3.3 kΩ resistance. It is now necessary to specify the current flowing in the various parts of the circuit and to insert the polarities of the voltage drops across the resistances (Fig. 8-3). In most cases the direction of the electron flow is obvious but sometimes the direction is doubtful. This does not present any difficulty since if the wrong direction is chosen, the analysis will show that the value of the current is negative. To write down the equation for a loop using Kirchhoff's Voltage Law, the convention is to move clockwise round the loop starting at any point and regard the voltage as positive if the negative polarity is first encountered. Kirchhoff's Votlage Law states that the algebraic sum of the voltages around the loop is zero so that the loop equations are:

Loop ABEF (starting at A).
$$- 3.3 I_2 + 19.8 - 12 + 4.7 I_1 = 0$$
$$4.7 I_1 - 3.3 I_2 = - 7.8$$

The units of the currents are milliamperes.

Loop BCDE (starting at E).
$$- 19.8 + 3.3 I_2 + 10 I_L = 0.$$
$$3.3 I_2 + 10 I_L = 19.8$$

Loop ABCDEF (starting at C).
$$10 I_L - 12 + 4.7 I_1 = 0.$$
$$4.7 I_1 + 10 I_L = 12.$$

The third loop equation does not provide additional information since it is obtained by adding the two preceding equations. Therefore, any two of the three loop equations are sufficient for the analysis.

In using Kirchhoff's Current Law, the convention is to regard currents entering a particular point as positive. Therefore:

$$I_1 + I_2 - I_L = 0$$

$$\text{or } I_2 = I_L - I_1$$

Substituting for I_L in the equation for loop BCDE:

$$3.3(I_L - I_1) + 10I_L = 19.8$$

$$-3.3I_L + 13.3I_L = 19.8$$

Multiplying the equation for loop ABCDEF by 3.3 and the preceding equation by 4.7:

$$15.51I_1 + 33I_L = 39.6$$

$$-15.51I_1 + 62.51I_L = 93.06$$

Adding these two equations

$$\text{Load current, } I_L = \frac{132.66}{92.51} = 1.39 \text{ mA.}$$

Method 3: Mesh Current Analysis.

In this method the convention is to ascribe clockwise mesh currents i_1, i_2 (electron flow) to two of the loops. In contrast to an analysis by Kirchhoff's Laws, an individual component can carry one or more mesh currents. In Fig. 8-3 the total current through the 3.3 kΩ resistance is the algebraic sum of i_1 and i_2. The mesh equations will be:

$$4.7i_1 + 3.3(i_1 - i_2) + 19.8 - 12 = 0.$$

$$8i_1 - 3.3i_2 = -7.8$$

$$\text{and } 3.3(i_2 - i_1) + 10i_2 - 19.8 = 0$$

$$-3.3i_1 + 13.3i_2 = 19.8.$$

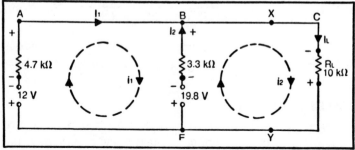

Fig. 8-3. Solution by Kirchhoff's laws.

Multiplying these two equations by 3.3 and 8 respectively:
$$26.4\,i_1 - 10.89\,i_2 = -25.74$$
$$-26.4\,i_1 + 106.4\,i_2 = 158.4$$
Then: $(106.4 - 10.89)i_2 = 158.4 - 25.74$
$$95.51\,i_2 = 132.66.$$

Load current: $i_2 = \dfrac{132.66}{95.51} = 1.39\,mA$.

Method 4: Superposition.

Step 1 Short out the 12 V constant EMF (Fig. 8-4A). By the current division rule, the value of I_{L1}, of the load current due to the 6 mA generator is $6\,mA \times \dfrac{R_T}{10\,k\Omega}$ where R_T is the total equivalent resistance of 3.3 kΩ, 4.7 kΩ and 10Ω in parallel.

Therefore $R_T = \dfrac{1}{\dfrac{1}{3.3} + \dfrac{1}{4.7} + \dfrac{1}{10}}$

$$= \frac{1}{0.303 + 0.213 + 0.1}$$

$$= \frac{1}{0.616} = 1.62\,k\Omega.$$

and $I_{L1} = 6\,mA \times 1.62\,k\Omega/10\,k\Omega = 0.972\,mA.$

Step 2 Open-circuit (remove) the 6 mA current generator (Fig. 8-4B). The total resistance of the remaining circuit is

$4.7\,k\Omega + \dfrac{3.3\,k\Omega \times 10\,k\Omega}{3.3\,k\Omega + 10\,k\Omega} = 7.10\,k\Omega.$ The load current due to the

12 V source is: $I_{L2} = \dfrac{12\,V}{7.18\,k\Omega} \times \dfrac{3.3\,k\Omega}{10\,k\Omega + 3.3\,k\Omega} = 0.415$ mA.

Since I_{L1} and I_{L2} are flowing in the same direction, the total load current by superposition is $0.974 + 0.415 = 1.39\,mA$, rounded off.

THEVENIN'S THEOREM

The current in a load resistance connected between two terminals X, Y of a network of resistances and generators is the same as if this load resistance were connected to a simple constant

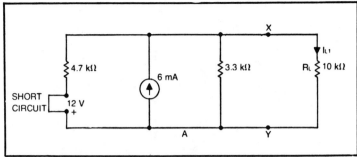

Fig. 8-4 A. Solution by superposition theorem, step one.

voltage generator whose EMF is the open-circuit voltage measured between X and Y and whose internal resistance is the resistance of the network looking back into the terminals X, Y with all generators replaced by resistances equal to their internal resistances; this last step would be achieved by short-circuiting all sources of constant voltage and open-circuiting all sources of constant current.

Figure 8-5 is used to illustrate *Thévenin's Theorem.* The Thévenin Voltage, E_{TH}, is the open-circuit voltage between the terminals X, Y with R_L disconnected, R_{TH} is the Thévenin resistance measured between X and Y with all generators replaced by resistance equal to their internal resistances.

Example 8-2

In the circuit of Fig. 8-1, calculate the value of I_L, using Thévenin's Theorem.

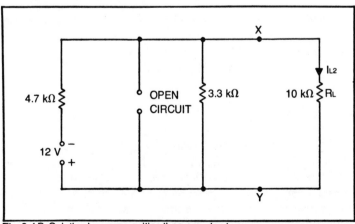

Fig. 8-4 B. Solution by superposition theorem, step two.

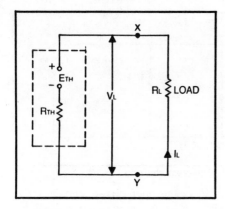

Fig. 8-5. Thevenin equivalent circuit.

Solution

Step 1: Remove the load R_L and convert the constant current generator into its constant voltage equivalent with an EMF equal to $6\,\text{mA} \times 3.3\,\text{k}\Omega = 19.8\,\text{V}$ (Fig. 8-6A).

The current I is $\dfrac{19.8\,\text{V} - 12\,\text{V}}{4.7\,\text{k}\Omega + 3.3\,\text{k}\Omega} = \dfrac{7.8}{8} = 0.975\,\text{mA}.$

Then $E_{OC} = E_{TH} = 19.8\,\text{V} - (3.3\,\text{k}\Omega \times 0.975\,\text{mA}) = 16.58\,\text{V}.$

Step 2: Short circuit the 12 V and 19.8 V constant voltage sources so that $R_{XY} = \dfrac{4.7 \times 3.3}{4.7\ \ 3.3} = 1.94\,\text{k}\Omega$ (Fig. 8-6B). Then the Thevenin equivalent circuit is as shown in Fig. 8-6 C.

Step 3: Reconnect the load R_L between X and Y. The load current is: $I_L \quad \dfrac{16.58\,\text{V}}{1.94\,\text{k}\Omega + 10\,\text{k}\Omega} = 1.39\,\text{mA}.$

NORTON'S THEOREM

The current in a load resistance connected between two terminals X, Y of a network consisting of generators and resistances is the same as if the load resistance were connected to a constant-current generator whose generated current, I_N, is equal to short-circuit current measured between X and Y. This constant-current generator has infinite internal resistance, but is

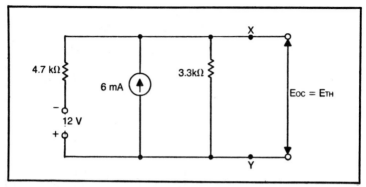

Fig. 8-6 A. Solution by Thevenin's theorem, step one.

placed in parallel with a resistance, R_N, equal to the resistance of the network looking back into the terminals X and Y with all generators replaced by resistances equal to their internal resistances (Fig. 8-7). This last step would be achieved by short-circuiting all sources of constant voltage and open-circuiting all sources of constant current.

This theorem is similar to Thévenin's theorem in that it enables a complicated network to be replaced by a single generator and a resistance. In this case, however, the generator is of the constant-current type and the resistance is in parallel, while in the case of Thévenin's Theorem, the equivalent generator is of the constant-voltage variety with the resistance in series. Since the procedure for finding R_{XY} is the same for both theorems, $R_N = R_{TH}$. The equivalent circuits are effectively identical to one another and may be transposed by using the results found in Chapter 4.

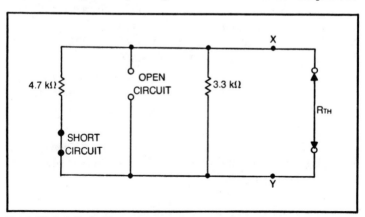

Fig. 8-6 B. Solution by Thevenin's theorem, step two.

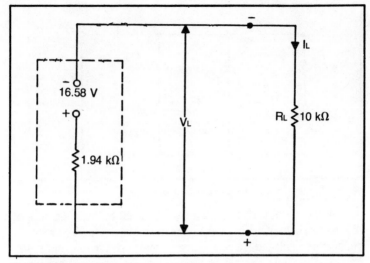

Fig. 8-6 C. The Thevenin equivalent circuit.

Therefore: $I_N = \dfrac{E_{TH}}{R_{TH}}$ and $E_{TH} = I_N \times R_N$.

Example 8-3

In the circuit of Fig. 8-1, calculate the value of I_L, using Norton's Theorem.

Solution

Step 1: Remove the load R_L and replace it by a short-circuit. Then convert the constant-voltage generator into its constant-current equivalent with a value of 12 V/4.7 kΩ = 2.55 mA, (Fig. 8-8 A). Then $I_N = 2.55\,mA + 6\,mA = 8.55\,mA$.

Step 2: Open-circuit the constant current sources (Fig. 8-8B). The $R_{xy} = R_N = 1.94$ kΩ, which is equal in value to the series Thevenin resistance. The equivalent Norton generator is shown in Fig. 8-8C.

Step 3: Remove the short circuit between X and Y and replace the load. Using the constant division rule in the Norton equivalent circuit, the load current is:

$$I_L = 8.55\,mA \times \frac{1.94\,k\Omega}{1.94\,k\Omega + 10\,k\Omega} = 1.39\,mA.$$

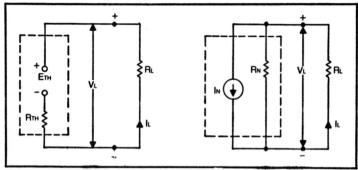

Fig. 8-7. Equivalent Thevenin and Norton circuits.

NODAL ANALYSIS

A node is a junction point in an electrical circuit. This method of analysis consists of using Kirchhoff's Current Law to obtain the equations for a particular node. Its application can be illustrated by again using the example of Fig. 8-1. The formal method of nodal analysis requires that only current sources are included; the circuit has therefore been redrawn as in Fig. 8-9 with the constant voltage generator replaced by its constant current equivalent. A ground has been included to provide a convenient reference node.

The purpose of the analysis is to find the value of V_L. As a convention it is assumed that the node voltage is negative with respect to the reference node; the actual polarity will be indicated by the sign of the answer. As the result of this convention electron current must flow away from the node through the resistors attached to the node. Sources which then force current into the

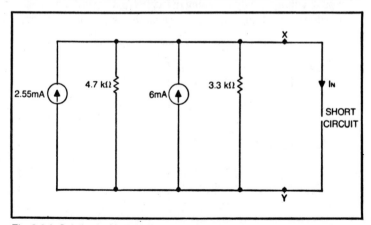

Fig. 8-8 A. Solution by Norton's theorem, step one.

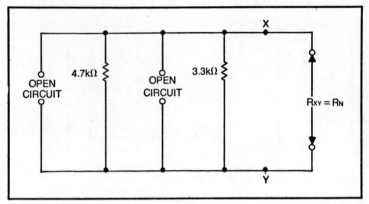

Fig. 8-8 B. Solution by Norton's theorem, step two.

node are considered to be positive while those driving current out of the node will be negative. The equation will then be: Algebraic sum of the source currents entering the node = total of the currents leaving the node through the resistors.

Therefore at the node N:

$$2.55 + 6.0 = \frac{V_L}{4.7} + \frac{V_L}{10} + \frac{V_L}{3.3}$$

$$V_L \times 0.6158 = 8.55$$

$$V_L = 1.39V \text{ and } I_L = \frac{13.9V}{10k\Omega} = 1.39\,mA.$$

Nodal voltage analysis using Kirchhoff's Current Law was able to solve the problem with a single equation while mesh current

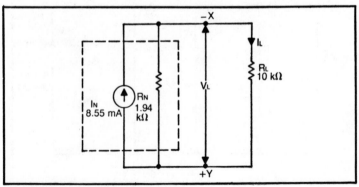

Fig. 8-8 C. The Norton equivalent circuit.

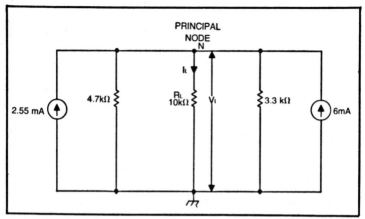

Fig. 8-9. Solution by nodal analysis.

analysis based on Kirchhoff's Voltage Law required two simultaneous equations. However the choice of a particular method depends on the circuit to be analyzed.

DELTA Y AND Pi TRANSFORMATIONS

The arrangements of resistors shown in Fig. 8-10 A is known as a delta connection (Δ is the Greek capital letter, delta). The resistors may be rearranged as in Fig. 8-10 B so that the formation resembles the Greek letter, P_i, π. Delta and P_i are therefore alternative names for the same arrangement of components.

Figure 8-11 A shows another resistor arrangement which is referred to as a wye (Y, connection). In Fig. 8-11 B the same arrangement has been modified to look like the letter tee (T). Consequently, either Wye or Tee may be used to signify the same component arrangement.

In many cases it is impossible to analyze a circuit directly in terms of series, parallel or series-parallel arrangements. However

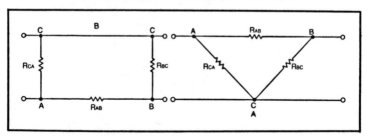

Fig. 8-10. Delta and Pi arrangements.

141

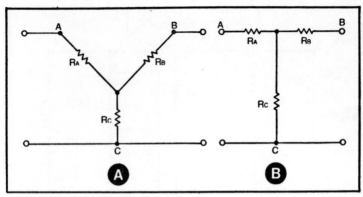

Fig. 8-11. Wye and tee arrangements.

if a delta arrangement of resistors can be replaced by an equivalent Y connection (or vice - versa), the resultant circuit may be changed into a simple series-parallel type.

In order that the delta and wye connections shall be equivalent, it is necessary that the resistances between the points A, B and C in both arrangements shall be the same. For example, in Fig. 8-10 A, using the product-over-sum formula: total resistance be-

tween   A and B = $\dfrac{R_{AB}(R_{BC}+R_{CA})}{R_{AB}+R_{BC}+R_{CA}}$ while in Fig. 8-11 A: total

resistance between A and B = $R_A + R_B$.

Therefore: $R_A + R_B = \dfrac{R_{AB}(R_{BC}+R_{CA})}{R_{AB}+R_{BC}+R_{CA}}$

Two other equations may be obtained for the total resistances between the points B, C and the points C, A.

Results derived from these three equations are:

$$R_A = \dfrac{R_{AB}R_{CA}}{R_{AB}+R_{BC}+R_{CA}}$$

$$R_B = \dfrac{R_{BC}R_{AB}}{R_{AB}+R_{BC}+R_{CA}}$$

$$R_C = \dfrac{R_{CA}R_{BC}}{R_{AB}+R_{BC}+R_{CA}}$$

These three equations would be used in transposing from a delta to a wye connection. The next three equations would be required for a wye to delta transformation:

$$R_{AB} = \frac{R_A R_B + R_B R_C + R_C R_A}{R_C}$$

$$R_{BC} = \frac{R_A R_B + R_B R_C + R_C R_A}{R_A}$$

$$R_{CA} = \frac{R_A R_B + R_B R_C + R_C R_A}{R_B}$$

A good case for using a delta - wye transformation is in the analysis of a bridge circuit (Fig. 8-12).

As already pointed out, the center 5.6 kΩ resistor is not directly connected in either series or parallel with the other resistors. However, if the points X, Y and Z are regarded as the corners of a delta formation, the equivalent Y connection will result in the circuit of Fig. 8-13. This new circuit is a simple series-parallel arrangement which may readily be analyzed. For the delta-wye transformation using the preceding equations:

$$R_X = \frac{5.6 \times 6.8}{5.6 + 6.8 + 3.3} = 2.425 \text{ k}\Omega$$

$$R_Y = \frac{5.6 \times 3.3}{5.6 + 6.8 + 3.3} = 1.177 \text{ k}\Omega$$

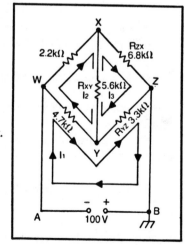

Fig. 8-12. Analysis of bridge circuit.

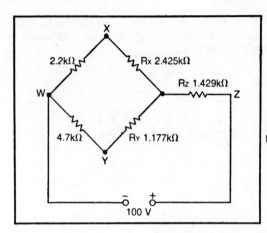

Fig. 8-13. Delta to wye transformation.

$$R_z = \frac{6.8 \times 3.3}{5.6 + 6.8 + 3.3} = 1.429 \text{ k}\Omega$$

In Fig. 8-13 the circuit's total resistance is:

$$R_T = 1.429 + \frac{(2.2 + 2.425) \times (4.7 + 1.177)}{(2.2 + 2.425) + (4.7 + 1.177)} = 4.017 \text{ k}\Omega.$$

$$I_T = \frac{100 \text{ V}}{4.017 \text{ k}\Omega} = 24.89 \text{ mA}.$$

The current through the 2.2 kΩ resistor is

$$24.89 \text{ mA} \times \frac{5.877 \text{ k}\Omega}{4.625 \text{ k}\Omega + 5.877 \text{ k}\Omega} = 13.93 \text{ mA}$$

The current through the 4.7 kΩ resistor is

$$24.89 - 13.93 = 10.96 \text{ mA}$$

The potential at point X is $-(100 - 13.93 \times 2.2) = -69.35$ V, while the potential at point Y is $-(100 - 4.7 \times 10.96) = 8\ 48.49$ V.

The current through the 5.6 kΩ resistor is $\frac{69.35 - 48.49}{5.6} = 3.73\ mA.$

The direction of the electron flow is from point X to point Y.

For comparison purposes, the current through the 5.6 kΩ resistor will be found by other methods.

Method 1: Thevenin's Theorem.

With this method the 5.6 Ω resistor is regarded as the load and therefore the first step is to remove this resistor from the circuit,

(Fig. 8-14 A). By the voltage division rule the potential at point X is

$$- 100 \text{ V} \times \frac{6.8 \text{ k}\Omega}{2.2 \text{ k}\Omega + 6.8 \text{ k}\Omega} = - 75.56 \text{ V.}$$ while the potential at

point Y is $- 100 \text{ V} \times \dfrac{3.3 \text{ k}\Omega}{4.7 \text{ k}\Omega + 3.3 \text{ k}\Omega} = - 41.25 \text{ V.}$ Then E_{TH} is

$75.56 - 41.25 = 34.31 \text{ V.}$

The second step is to short out the 100 V source and then obtain the open circuit resistance between the points X and Y, (Fig. 8-14 B). With respect to these points the 2.2 kΩ, 6.8 kΩ resistors are in parallel; the 4.7 kΩ, 3.3 kΩ resistors are also in parallel but the two parallel combinations are in series so that R_{TH} is:

$$\frac{2.2 \times 6.8}{2.2 + 6.8} + \frac{3.3 \times 4.7}{3.3 + 4.7} = 1.66 + 1.94 = 3.60 \text{ k}\Omega.$$

The Thévenin equivalent circuit is shown in Fig. 8-14 C. When the 5.6 kΩ resistor is replaced, the load current is

$$\frac{34.31 \text{ V}}{3.6 \text{ k}\Omega + 5.6 \text{ k}\Omega} = 3.73 \text{ mA.}$$

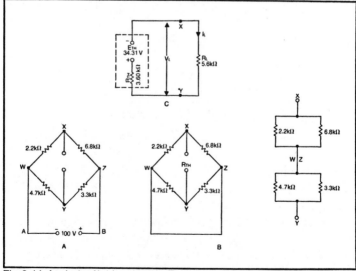

Fig. 8-14. Analysis of bridge circuit by Thevenin's theorem.

145

Method 2: Norton's Theorem.

The first step is to remove the 5.6 kΩ load and replace the resistor by a short circuit (Fig. 8-15 A). The total resistance of the circuit is

$$\frac{2.2 \times 4.7}{2.2 + 4.7} + \frac{6.8 \times 3.3}{6.8 + 3.3} = 1.499 + 2.222 = 3.721 \text{ k}\Omega.$$

Using the current division rule, the current in the 2.2 kΩ resistor is $\frac{100 \text{ V}}{3.721 \text{ k}\Omega} \times \frac{4.7 \text{ k}\Omega}{2.2 \text{ k}\Omega + 4.7 \text{ k}\Omega} = 18.31$ mA, while

the current in the 6.8 kΩ resistor is $\frac{100 \text{ V}}{3.721 \text{ k}\Omega} \times \frac{3.3 \text{ k}\Omega}{3.3 \text{ k}\Omega + 6.8 \text{ k}\Omega}$

$= 8.78$ mA. Therefore: I_N is $18.31 - 8.78 = 9.53$ mA and since $R_N = R_{TH}$ (Fig. 8-15 B) the equivalent Norton circuit is as shown in Fig. 8-15C. When the 5.6 kΩ resistor is replaced, its current by the

division rule is 9.53 mA $\times \frac{3.6 \text{ k}\Omega}{5.6 \text{ k}\Omega + 3.6 \text{ k}\Omega} = 3.73$ *mA.*

Method 3: Mesh Current Analysis. Referring to Fig. 8-12 the mesh equations are:

Mesh AWYZB $(I_1 - I_2) 4.7 + (I_1 - I_3) 3.3 = 100.$

Mesh WXY $2.2 I_2 + 5.6 (I_2 - I_3) + 4.7 (I_2 - I_1) = 0.$

Mesh ZYX $3.3 (I_3 - I_1) + 5.6$ '$(I_3 - I_2) + 6.8 I_3 = 0.$

Therefore:

$$8 I_1 - 4.7 I_2 - 3.3 I_1 = 100.$$

$$- 4.7 I_1 + 12.5 I_2 - 5.6 I_3 = 0.$$

$$- 3.3 I_1 - 5.6 I_2 + 15.7 I_3 = 0.$$

The solutions to these equations are: $I_1 = 24.89$ mA, $I_2 = 13.93$ mA, $I = 10.20$ mA. The current through the 5.6 kΩ resistor is: $I_t - I_3 = 13.93 - 10.20 = 3.73$ *mA.*

In this case nodal analysis would have no advantage over mesh analysis since there are three principal nodes at W, X, Y and therefore the number of simultaneous equations would be the same for both methods.

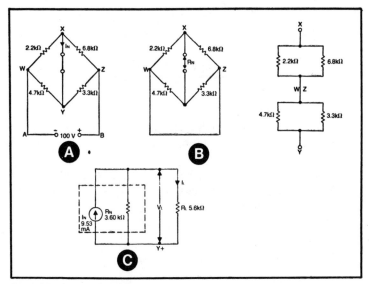

Fig. 8-15. Analysis of bridge circuit by Norton's theorem.

CHAPTER SUMMARY

☐ *Kirchhoff's Voltage Law.*

The algebraic sum of the voltages around a closed loop (or mesh) is zero.

☐ *Kirchhoff's Current Law.*

The algebraic sum of the currents at a junction point (or node) is zero.

☐ *Thévenin's Theorem.*

Thévenin voltage, E_{TH} = open circuit voltage between the load terminals (load removed).

Thévenin Resistance, R_{TH} = open circuit resistance between the load terminals (load removed) with all sources replaced by their internal resistances.

☐ *Norton's Theorem.*

Norton current, I_N = short-circuit current between the load terminals (load removed).

Norton resistance, $R_N = R_{TH}$.

☐ *Nodal Analysis.*

At a principal node the algebraic sum of the source currents = the sum of the currents through the resistors.

☐ *Delta to Wye Transformation.*

$$R_A = \frac{R_{AB} \, R_{CA}}{R_{AB} + R_{BC} + R_{CA}} \quad \text{etc.}$$

$$R_{AB} = \frac{R_A R_B + R_B R_C + R_C R_A}{R_C} \quad \text{etc.}$$

☐ *Millman's Theorem.*

Total current of the equivalent constant-current generator = algebraic sum of the currents in the individual constant current generators.

Internal resistance of the equivalent constant current generator = total resistance of the constant current generators' internal resistances in parallel.

Introduction to Alternating Current

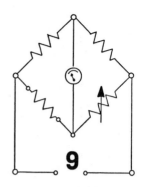

9

Previous Chapters have been concerned with direct current in which the electron flow has always been in the same direction; however the magnitude of the flow was not necessarily constant.

COMPARISON BETWEEN DIRECT CURRENT AND ALTERNATING CURRENT

With alternating current, the current reverses direction periodically and has an average value of zero. The current may assume a variety of waveforms some of which are shown in Fig. 9-1. In each case a single complete waveform is known as a cycle and the number of cycles occuring in one second is the frequency, f. This is the rate at which the waveform repeats and is measured in cycles per second (cs) or hertz (Hz). The time interval taken by a complete cycle is the period, T, which is equal to the reciprocal of

the frequency. Therefore: $T \text{ (seconds)} = \dfrac{1}{f(\text{Hz})}$ and $f = \dfrac{1}{T}$ Hz.

If the average value of the waveform is not zero, it is regarded as a combination of a DC component and an AC component. The 60 Hz commercial AC line voltage has a sine waveform whose shape is shown in Fig. 9-2. This curve is the result of plotting the mathematical sine function, $\text{Sin}\,\theta$, against the angle, θ, measured in either degrees or radians. The radian is the angle at the center of a circle which subtends (is opposite to) an arc equal in length to the radius. Since the total length of the circumference is $2\pi \times$ radius, the angle of 360° must be equivalent to 2π radians. Therefore:

$$\theta \text{ (degrees)} = \frac{360}{2\pi} \times \theta \text{ (radians)} = \frac{180}{\pi} \times \theta \text{ (radians)}$$

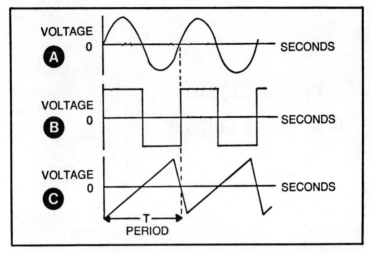

Fig. 9-1. Examples of alternating voltage waveforms: A, sine wave, B, square wave, C, sawtooth wave.

and θ (radians) $= \dfrac{\pi}{180} \times \theta$ (degrees).

Therefore: 1 radian $= \dfrac{180}{\pi}$ degrees $= 57.296° = 57° \, 17' \, 45''$.

Sine wave AC is of particular importance since Fourier analysis allows the waveforms in Fig. 9-1 to be broken down in a series of sine waves which consist of a fundamental component and harmonics (Fig. 9-3 and Appendix B). The fundamental freuqency, f, is the same as the waveform's repetition rate, the second harmonic has a frequency 2f, the third harmonic 3f etc, (Fig. 9-4). Compared with other AC waveforms the sine wave also has a high form factor.

An angular velocity of rotation is measured in radians per second and is normally designated by the Greek letter, ω – omega. If the angle of rotation is θ radians. $\theta = \omega t$ where t is the time in seconds.

During a cycle, a total rotation of 2π radians occurs in the time of T seconds. Then:

$$\omega = \frac{\theta}{t} \quad = \quad \frac{2\pi}{T} \quad = 2\pi f \text{ radians per second.}$$

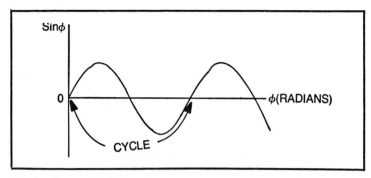

Fig. 9-2. The mathematical sine wave.

THE SIMPLE AC GENERATOR (ALTERNATOR)

The AC generator is a machine which is capable of converting mechanical energy into alternating electrical energy. Figure 9-5 A shows a shaft, D, which is driven around by mechanical energy. Attached to the shaft but insulated from it is a loop which rotates between the poles, P, of a permanent magnet or an electromagnet. The ends of the loop are joined to two slip rings, s, s' which as they rotate, make contact with stationary carbon brushes, C, C'; these brushes are then connected to the electrical load.

The pole pieces, PP' are especially shaped to provide a constant flux density in which the loop rotates as shown in Fig. 9-5 B.

The loop of N turns moves with an angular velocity of ω radians per second so that the velocity of each of the conductors, A,

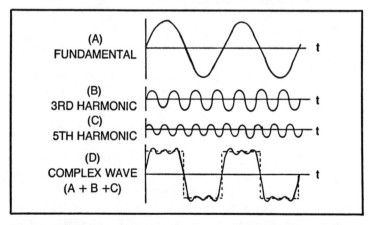

Fig. 9-3. Synthesis of an approximate square wave from a fundamental sine wave and its odd harmonics.

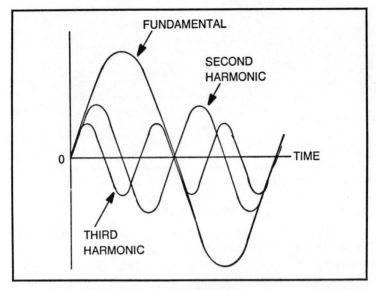

Fig. 9-4. Fundamental sine wave and its harmonics.

B, is $v = \omega b$ meters per second where 2b is the width of the loop in meters. The component of the velocity at right angles to the magnetic flux is $v \sin \phi$. Using the equations from Chapter five, the total voltage induced in the loop is: $e = 2BLNv \sin \phi$

$$= BAN\omega \sin \omega t \text{ volts}$$

where B = constant flux density in teslas

    L = length of each conductor in meters

    A = 2Lb = area of each turn in the loop

    t = time in seconds.

Therefore the instantaneous output from the generator is: $e = E_{max}$ $\sin \omega t$ where $E_{max} = BAN\omega$ = the maximum or peak value of the output voltage.

The waveform of this output voltage is shown in Fig. 9-6. Since the sine wave is symmetrical about the zero line, the peak-to-peak value (E_{p-p}) is equal to twice the peak value of $2E_{max}$. It is an AC voltage's peak-to-peak value which is normally measured on an oscilloscope display.

For the two pole machine of Fig. 9-5 A, one complete rotation of the loop will generate one cycle of AC through the load. If the output frequency is 60 Hz, the loops speed of rotation is $60 \times 60 = 3600$ revolutions per minute (rpm); however if the alternator has four poles, one complete rotation will generate two cycles of AC and therefore the speed required to produce a 60 Hz voltage is only

152

1800 rpm. The frequency:

$$f = \frac{Np}{60} \text{ Hz and } N = \frac{60f}{p} \text{ rpm.}$$

where p = number of pairs of poles
N = speed of rotation in rpm.

Example 9-1

A coil of 1000 turns is rotated at 3600 rpm in a magnetic field having a uniform flux of 0.06 tesla. The average area of each turn is 50 cm² and the axis of rotation is at right angles to the direction of the flux. Calculate (a) the frequency (b) its period (c) the angular velocity and (d) the maximum value of the generated voltage. Write down a trigonometrical expression for the instantaneous voltage and calculate its value when the coil has rotated 7 radians from the position of zero voltage.

Solution

(a) Frequency, $f = \frac{3600}{60} = 60\,Hz.$

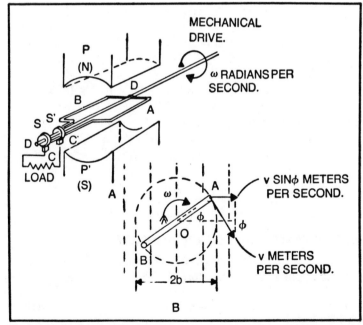

Fig. 9-5. Basic as generator.

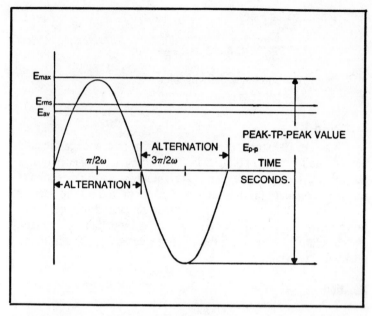

Fig. 9-6. Values associated with the sine wave.

(b) Period, $T = \dfrac{1}{f} = \dfrac{1}{60}$ second $= \mathit{16.67\ milliseconds}$.

(c) Angular velocity, $\omega = 2\pi f = 2\pi \times 60 = \mathit{377}$ radians per second.

(d) Maximum voltage, $E_{max} = BAN\omega$
$$= 0.06 \times 50 \times 10^{-4} \times 10^{3} \times 377$$
$$= 113\,V.$$

Therefore the instantaneous voltage, $e = E_{max} \sin \omega t$
$$= 113 \sin 377\,t.$$
$$= 113 \sin\phi$$

If $\phi = 7$ radians,
$$e = 113 \sin (7\ radians)$$
$$= \mathit{74\ V}.$$

EFFECTIVE OR ROOT-MEAN-SQUARE (rms) VALUE

In Fig. 9-7A an AC voltage of instantaneous voltage $e = E_{max} \sin\omega t$ is applied across resistor, R. The instantaneous power

dissipated is: $p = \dfrac{e^2}{R} = \dfrac{E^2_{max} \sin^2\omega t}{R} = \dfrac{E^2_{max}(1 - \cos 2\omega t)}{2R}$

154

The instantaneous power ranges from zero to a peak value of $\frac{E^2_{max}}{R}$. The average value of $\cos 2\omega t$ over a complete cycle is zero; this is indicated by the symmetry of the instantaneous power curve in Fig. 9-7B.

The average power is therefore: $P_{av} = \frac{E^2_{max}}{2R} = \left(\frac{\frac{E_{max}}{\sqrt{2}}}{R}\right)^2$

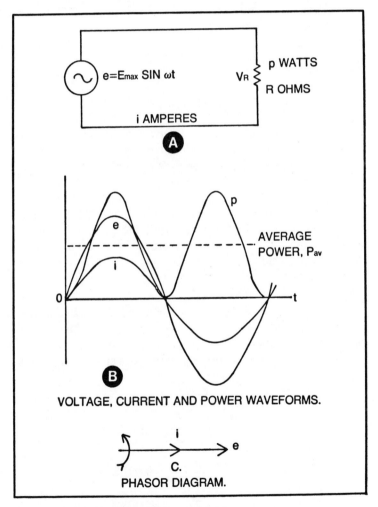

Fig. 9-7. Resistance in ac circuits.

155

The quantity $\sqrt{\dfrac{E_{max}}{2R}} = \dfrac{E_{max}}{1.414} = 0.707 \; E_{max}$ is called the

Root-Mean-Square (rms) value of the AC sinewave voltage. In other words, E_{rms} is equal to the square root of the mean value of e^2. The rms value of an alternating voltage is therefore the steady or constant voltage which would provide the same mean power or heating effect as the alternating voltage. To convert between peak and rms value (Fig. 9-6) E_{peak} or $E_{max} = \sqrt{2} \times E_{rms} = 1.414 \times E_{rms}$

and $E_{rms} = \dfrac{E^{peak}}{\sqrt{2}} = \dfrac{E^{peak}}{1.414} = 0.707 \times E_{peak}$.

Similarly: $I_{rms} = \dfrac{I_{peak}}{\sqrt{2}}$ etc.

Most alternating voltage and current measurements are in terms of rms values. However insulation in AC circuits must normally be able to withstand the peak voltage. For example, the commercial 110 V, 60 Hz line voltage has a peak value of $110 \times \sqrt{2}$ or approximately 155 V.

If the waveform of the current flowing through a resistor contains both DC and AC components, the rms value of the complete current is $\sqrt{I^2_{DC} + I^2_{rms}}$ where I_{DC} is the DC component and I_{rms} is the effective value of the AC component (Fig. 9-8).

AVERAGE VALUE AND FORM FACTOR

The full cycle of a sine wave is composed of two alternations; one alternation extends from 0° to 180° and the other from 180° to 360° (Fig. 9-6). Although the average value of the sine wave over the complete cycle is zero, the average value over an alternation is:

$$E_{av} = \dfrac{\displaystyle\int_o^{\frac{\pi}{\omega}} E_{max} \, Sin\omega \; dt}{\displaystyle\int_o^{\frac{\pi}{\omega}} dt}$$

$$= \dfrac{2}{\pi} E_{max} = 0.637 \, E_{max}.$$

The average value over an alternation is shown in Fig. 9-6.

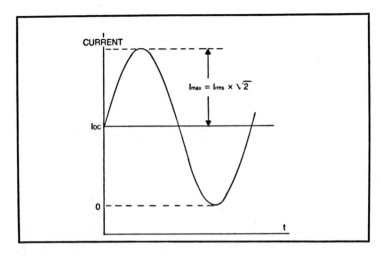

Fig. 9-8. Current containing dc and ac components.

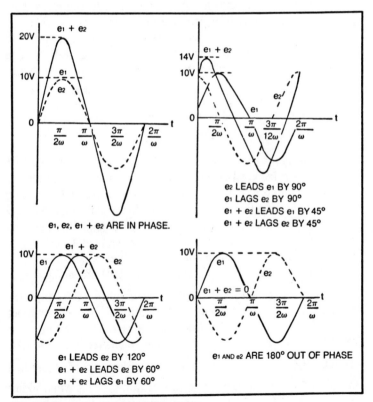

e_1, e_2, $e_1 + e_2$ ARE IN PHASE.

e_2 LEADS e_1 BY 90°
e_1 LAGS e_2 BY 90°
$e_1 + e_2$ LEADS e_1 BY 45°
$e_1 + e_2$ LAGS e_2 BY 45°

e_1 LEADS e_2 BY 120°
$e_1 + e_2$ LEADS e_2 BY 60°
$e_1 + e_2$ LAGS e_1 BY 60°

e_1 AND e_2 ARE 180° OUT OF PHASE

Fig. 9-9. Examples of phase relationships.

The ratio of the rms value to the average value of an alternating current or voltage is called the "form factor" of the waveform. For the sine wave:,

$$\text{Form factor} = \frac{0.707\,E_{max}}{0.637\,E_{max}} = 1.11$$

Compared with other ac waveforms the sine wave is relatively easy to generate and has a superior form factor.

PHASE

If two DC voltages are in series, they are either aiding or opposing so that only addition or subtraction is necessary in order to obtain the total voltage. However two AC waveforms of the same frequency may not reach similar points in the cycle, for example peak or zero values, at the same time. The amount by which the two waveforms are out of step is referred to as their phase difference which is measured in either degrees or radians.

In Fig. 9-10A the e_2 waveform reaches its peak, x, earlier in time than the e, waveform with its corresponding peak, y. The phase difference between the two waveforms is ϕ radians with e_2 leading e (or e_1 lagging e_2). The waveform equations are:

$$e_2 = E_{2max}\,\text{Sin}\,(\omega t + \phi)$$

and

$$e_1 = E_{1max}\,\text{Sin}\,\omega t.$$

Similarly the equation $e_3 = E_{3max}\,\text{Sin}\,(\omega t - \phi)$ would mean that e_3 lags e_1 (or e_1 leads e_3) by ϕ radians; also e_2 leads e_3 by 2ϕ radians.

Phase differences may therefore be either leading or lagging and the angles usually extend up to 180°. In the particular case of ϕ = 180° or π radians, the terms "leading" or "lagging" are not used since 180° lagging would have the same meaning as 180° leading. In most cases there is little point in using angles greater than 180°, since, for example, 270° lagging is equivalent to 90° leading.

It is impossible to add together two or more AC voltages (or currents if some form of parallel circuit is involved) without knowing their magnitudes and their phase relationship. For example if a resistor and an inductor are connected in a series AC circuit, their voltages are 90° or $\frac{\pi}{2}$ radians out of phase and this phase relationship must be taken into account when adding the voltages. Figure 9-9 shows the result of

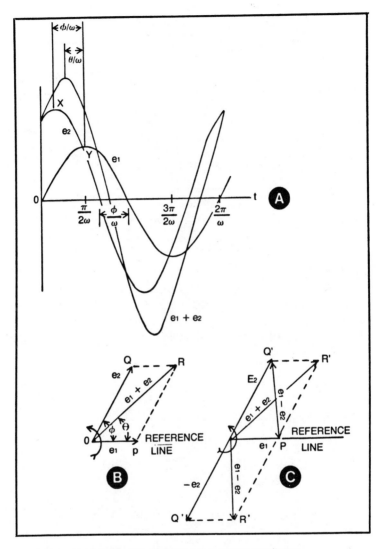

Fig. 9-10. Addition and subtraction of the two sine wave voltages.

adding two AC voltages, each of 10 volts peak, but with phase differences which are 0°, 90°, 120°, 180°; the corresponding resultant voltages have peak values of 20 V, 14.14 V, 10 V and zero.

In the general case of $e_1 = E_{1max} \sin\omega t$ and $e_2 = E_{max} \sin(\omega t + \phi)$, the sum of the two AC voltages is:

$$e_1 + e_2 = E_{1max} \sin\omega t + E_{2max} \sin(\omega t + \phi)$$

159

$$= E_{1max} \sin\omega t + E_{2max} \cos\phi \sin\omega t + E_{2max} \sin\phi \cos\omega t$$

$$= \sqrt{E^2_{1max} + E^2_{2max} + 2E_{1max} E_{2max} \cos\phi} \; \sin(\omega t + \phi)$$

where:

$$\phi = \text{Inv. (inverse) Tan. or Arc. tan. or } \tan^{-1}\left(\frac{(E_{2max} \sin\phi)}{E_{1max} + E_{2max} \cos\phi}\right)$$

The resultant voltage, $e_1 + e_2$, has therefore the same frequency as e_1 and e_2. It has a peak value of $\sqrt{E^2_{1max} + E^2_{2max} + 2E_{1max} E_{2max} \cos\phi}$ and leads e_1 by ϕ radians (Fig. 9-10A).

It follows that:

$$e_1 - e_2 = \sqrt{E^2_{1max} + E^2_{2max} - 2E_{1max} E_{2max} \cos\phi} \; \sin(\omega t + \theta)$$

where $\phi = $ Inv. Tan. $\left(\dfrac{E_{2max} \sin\phi}{E_{2max} \cos\phi - E_{1max}}\right)$

Example 9-2

A sinewave voltage of peak-to-peak value 24 V is applied across a 80 Ω resistor. Calculate (a) the peak, average (over an alternation) and effective values of the sinewave voltage and (b) the values of the peak power and the average power dissipated over the cycle.

Solution

(a) Peak value, $E_{max} = \dfrac{E_{p-p}}{2} = \dfrac{24}{2} = 12\,V.$

Average value, $E_{av} = \dfrac{2}{\pi} \times 12 = 0.637 \times 12 = 7.64\,V.$

Effective value, $E_{rms} = \dfrac{12}{\sqrt{2}} = 0.707 \times 12 = 8.49\,V.$

(b) Peak power $= \dfrac{E^2_{max}}{R} = \dfrac{12^2}{80} = \dfrac{144}{80} = 1.8\,W.$

Average power $= \dfrac{\text{peak power}}{2} = \dfrac{1.8}{2} = 0.9\,W.$

Alternatively, average power $= \dfrac{E^2_{rms}}{R} = \dfrac{8.49^2}{80} = 0.9\,W.$

Example 9-3

A sinewave voltage has an effective value of 110 V. Calculate its peak-to-peak value and its average value over an alternation.

Solution

$$\text{Peak-to-peak value} = \sqrt{2}\,E_{rms} = 2.828 \times E_{rms}$$
$$= 2.828 \times 110$$
$$= 311\,V.$$

$$\text{Average value over an alternation} = \frac{2\sqrt{2}}{\pi} \times E_{rms}$$
$$= 0.9 \times 110 = 99\,V.$$

Example 9-4

Two alternating voltages are represented by $e_1 = 10\,\text{Sin}\,\omega t$ and $\left(\omega t + \dfrac{\pi}{6}\right)$. What is the phase relationship between between e_1 and e_2? What are the trigonometrical expressions for $e_1 + e_2$ and $e_1 - e_2$? Verify these results by drawing to scale the sinewaves for $e_1, e_2, e_1, + e_2, e_1 - e_2$ and $e - e_1$.

Solution

Since $\dfrac{\pi}{6}$ radians $= 30°$, e_1 lags e_2 by $30°$ or e_2 leads e_1 by $30°$.

$$e_1 + e_2 = \sqrt{10^2 + 15^2 + 2 \times 10 \times 15 \times \text{Cos}\,30°}\,\text{Sin}\,(\omega t + \phi_1)$$

where $\phi_1 = \text{Inv. Tan.}\left(\dfrac{15\,\text{Sin}\,30°}{10 + 15\,\text{Cos}\,30°}\right) = 18.1°$

Therefore $e_1 + e_2 = 24.18\,\text{Sin}\,(\omega t + 18.1°)$ which leads e_1 by $18.1°$ but lags e_2 by $30° - 18.1° = 11.9°$.

$$e_1 - e_2 = \sqrt{10^2 + 15^2 - 2 \times 10 \times 15 \times \text{Cos}\,30°}\,\text{Sin}\,(\omega t + \phi_2)$$

where $\phi_2 = \text{Inv. Tan.}\left(\dfrac{15\,\text{Sin}\,30°}{15\,\text{Cos}\,30° - 10}\right) = -111.7°$

Therefore $e_1 - e_2 = 8.07\,\text{Sin}\,(\omega t - 111.7°)$ which leads e_1 by $68.3°$ and e_2 by $38.3°$.

Note that $e_2 - e_1 = 8.07$ Sin $(\omega t + 68.3°)$ which is $180°$ out of phase with $e_1 - e_2$.

The waveforms for e_1, e_2, $e_1 + e_2$ and $e_2 - e_1$, are shown in Fig. 9-11.

REPRESENTATION OF AN AC VOLTAGE OR CURRENT BY A ROTATING PHASOR

So far in this chapter an AC voltage or current has been represented either by a trigonometrical equation or graphically by a waveform. Both these methods tend to be tedious, particularly if voltages (or currents) have to be added or subtracted.

A third method involves phasor representation. If the equation of an AC voltage is $e = E_{max}$ Sinωt, its phasor is a line whose length is equal to the peak value, E_{max}. By convention the phasor is assumed to rotate in the positive or counter-clockwise direction with an angular velocity of ω radians per second. The vertical projection of the phasor on the horizontal reference line equals the instantaneous value of the AC voltage (Fig. 9-12). Therefore as the phasor rotates its extremity can be said to trace out the AC voltage's sinusoidal waveform with a frequency equal to the phasor's speed of rotation in revolutions per second. The phasor diagram therefore contains the same information as the waveform presentation and it is obviously easier to work with lines rather than sine waves.

Since there is a constant relationship between rms and peak values, (rms value = $0.707 \times$ peak value), the length of the phasor may be used to indicate the rms value.

If two sine waves are not in phase, the phase difference will be represented by the angular separation between their phasors. In Fig. 9-10 B, the phasors for $e_1 = E_{1max}$ Sin ωt and $e_2 = E_{2max}$ Sin $(\omega t + \phi)$ will be separated by an angle of ϕ radians with e_2 leading e_1 as the phasors rotate in the counter-clockwise direction. In many phasor diagrams one of the phasors, in this case e_1, is chosen as a reference and is placed along the horizontal line.

ADDITION AND SUBTRACTION OF PHASORS

Previous sections have shown how to add together two sine waves trigonometrically and by a waveform presentation. In order to add the phasors e_1 and e_2, it is necessary to construct the parallelogram of Fig. 9-10 B and then draw the diagonal OR to represent the phasor, $e_1 + e_2$. It may then be shown that the length, OR, is $\sqrt{E^2_{1max} + E^2_{2max} + 2E_{1max} E_{2max}}$ Cos ϕ and that the angle, ϕ,

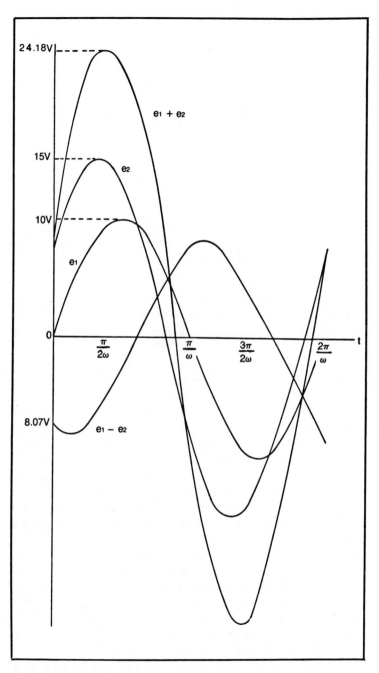

Fig. 9-11. Waveforms related to example 9-4.

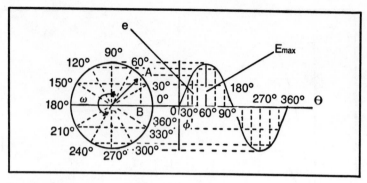

Fig. 9-12. Phasor representation of a sine wave.

$$\text{is Inv. (Arc) Tan.} \left(\frac{E_{2max} \, Sin \, \phi}{E_{1max} + E_{2max} \, Cos\phi} \right)$$

These were exactly the same results which were obtained by adding the sine waves trigonometrically and this therefore justifies the use of the parallelogram of phasors for the operation of addition.

To obtain $e_1 - e_2 = e_1 (-e_2)$ in the process of subtracting phasors, the method involves finding $(- e_2)$, which is then added to e_1. Since $- e_2 = - E_{2max} \, Sin(\omega t + \phi) = E_{2max} \, Sin \, (\omega t + \phi + \pi)$, the negative sign causes the phasor to be rotated through π radians or 180° as shown in Fig. 9-10 C. Adding the phasors e_1 and $- e_2$ together produces the diagonal OR′ which represents $e_1 - e_2$ and is also the same as the other diagonal of the parallelogram OPQR. In other words, one of the diagonals, OR, represents the phasor sum and the other diagonal, QP, is the phasor difference.

It is worth mentioning that prior to about 1960 it was common practice to refer to phasors as vectors. A vector is a quantity which possesses both magnitude and direction while a scalar possesses a magnitude only. Therefore force and velocity are vectors while power is a scalar. The vectors in mechanics obey the same rules as phasors for addition and subtraction but the rules for multiplication and division are totally different; to emphasize this distinction, the term "phasor" was introduced. To establish the rules for multiplication and division of phasors will require the use of complex algebra which is discussed in chapter 13.

Now that we have the phasor, waveform and trigonometrical representations of sine wave voltages and currents, we can examine the effects of resistance, inductance and capacitance in AC circuits.

Example 9-5

Two alternating voltages are represented by $e_1 = 10 \sin \omega t$ and $e_2 = 15 \sin \left(\omega t + \dfrac{\pi}{6} \right)$. Draw to scale a phasor diagram containing e_1, e_2, and construct the phasors representing $e_1 + e_2$ fi fi$_{fi}$ $- e_2$ and $e_2 - e_1$. Compare the results with the answers to the previous example.

Solution

The relevant phasor diagram is shown in Fig. 9-13.

Within the limits of measurement, the phasors are:

$$e_1 + e_2 = 24.2 \sin (\omega t + 18°)$$
$$e_1 - e_2 = 8.1 \sin (\omega t - 112°)$$
$$e_2 - e_1 = 8.1 \sin (\omega t + 68°)$$

RESISTANCE IN AN AC CIRCUIT

Let an instantaneous AC voltage, $e = E_{max} \sin \omega t$, be applied to a resistor, R (Fig. 9-7); note that $v_R = - E_{max} \sin \omega t$ and $e + v_R = 0$. Since Ohm's Law applied at any instant, the instantaneous

$$\text{current } i = \frac{e}{R} = \frac{E_{max}{}^{\sin \omega t}}{R}$$

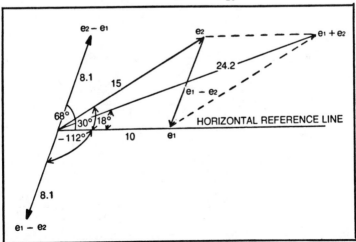

Fig. 9-13. Phasor diagram for example 9-5.

$= I_{max} \, \text{Sin} \, \omega t$, e and i are therefore in phase as shown in the waveform and phasor diagrams. Also:

$$\frac{E_{max}}{I_{max}} = \frac{1.414 \, E_{rms}}{1.414 \, I_{rms}} = \frac{E_{rms}}{I_{rms}} = R.$$

The instantaneous power, p, is:

$$p = ei = E_{max} I_{max} \, \text{Sin}^2 \, \omega t.$$

$$= \frac{E_{max} I_{max}}{2} \, (1 - \text{Cos} \, 2\omega t)$$

The average power over the cycle is:

$$\frac{E_{max} I_{max}}{2} = \frac{E_{max}}{\sqrt{2}} \times \frac{I_{max}}{\sqrt{2}} = E_{rms} \times I_{rms}.$$

The AC formulas for resistances in series and parallel are the same as those for DC.

Example 9-6

In Fig. 9-7, $e = 18 \, \text{Sin} \left(\omega t + \frac{\pi}{4} \right)$ volts and $R = 6.8 \, k\Omega$.

Calculate the effective values of the current, the peak power and the average power over the cycle. Write down the trigonometrical expression for the instantaneous current.

Solution

Effective voltage, $E_{rms} = \frac{E_{max}}{\sqrt{2}} = 18 \times 0.707 = 12.73 \, V.$

Effective current, $I_{rms} = \frac{E_{rms}}{R} = \frac{12.73 \, V}{6.8 \, k\Omega} = 1.87 \, mA.$

Peak power $= \frac{E^2_{max}}{R} = 47.6 \, mW.$

Average power $= E_{rms} \times I_{rms} = 23.8 \, mW.$

Since e and i are in phase, the instantaneous current, $i = \frac{18}{6.8}$

$\text{Sin} \left(\omega t + \frac{\pi}{4} \right) = 2.65 \, \text{Sin} \left(\omega t + \frac{\pi}{4} \right)$ mA.

INDUCTANCE IN AN AC CIRCUIT

Let an instantaneous AC voltage, $e = E_{max} \sin \omega t$, be applied to an inductor L (Fig. 9-14). The inductor is regarded as possessing inductance only and therefore its resistance is zero.

The induced voltage or counter EMF, $v_L = -L \times$ rate of change of current $= -L \dfrac{di}{dt}$, using calculus notation.

By Kirchhoff's Voltage Law, $e + v_L = 0$

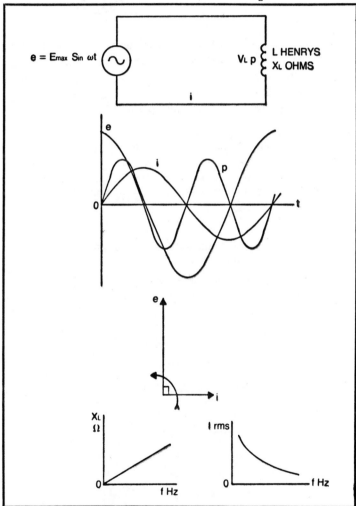

Fig. 9-14. Inductance in ac circuits.

and therefore: $e = L \dfrac{di}{dt}$ so that $\dfrac{di}{dt} = \dfrac{e}{L} = \dfrac{\text{max Sin } \omega t}{L}$.

Integrating with respect to t:

$$i = - \frac{E_{max} \text{ Cos } \omega t}{\omega L} = \frac{E_{max} \text{ Sin } (\omega t - \frac{\pi}{2})}{\omega L}$$

$$= I_{max} \text{ Sin } (\omega t - \frac{\pi}{2}).$$

i therefore lags e by $\dfrac{\pi}{2}$ radians or 90°. Alternatively e leads i by 90° and this may be remembered by the word "eLi". (for an inductor L, the instantaneous voltage, e, leads the instantaneous current, i, by 90°). When voltage leads current, the phase angle is considered to be positive and therefore $\phi = +90°$. Also:

$$\frac{E_{max}}{I_{max}} = \frac{E_{rms}}{I_{rms}} = \omega L = 2\pi \, fL = X_L \, \Omega.$$

Therefore $I_{rms} = \dfrac{E_{rms}}{X_L} = \dfrac{E_{rms}}{2\pi fL}$, $f = \dfrac{X_L}{2\pi LHz}$ and $L = \dfrac{X_L}{2\pi f}$ H

The quantity X_L is called the inductive reactance and measures the inductor's total opposition to alternating current.

Note that X_L is directly proportional to frequency while I_{rms} is inversely proportional to frequency. These relationships are illustrated in Fig. 9-14.

The instantaneous power in the circuit is:

$$p = ei = E_{max} \text{ Sin } \omega t \times I_{max} \text{ Sin } (\omega t - \frac{\pi}{2})$$

$$= \frac{E_{max} I_{max}}{2} \text{ Sin } 2(\omega t - \frac{\pi}{2})$$

$$= E_{rms} \times I_{rms} \text{ Sin } \left[2(\omega t - \frac{\pi}{2}) \right]$$

The mean value of Sin $[2(\omega t - \frac{\pi}{2})]$ is zero and therefore the average value of the power over the complete cycle is also zero (Fig. 9-14). Once again the frequency of the instantaneous power curve is a second harmonic. During two quarters of the cycle energy is taken from the source in establishing a magnetic flux around the inductor while during the remainder of the cycle, the magnetic flux collapses and energy is returned to the source. This illustrates the difference between resistance and reactance. Both resistance and reactance limit the flow of alternating current but while resistance dissipates power, reactance does not.

The product $E_{rms} I_{rms}$ is called the idle, wattless or reactive power which is measured in volt-amperes-reactive (VArs).

Now consider inductive reactances in series. The equation for inductors in series is: $L_T = L_1 + L_2 + L_3 + \ldots\ldots + L_N$
Therefore: $2\pi fL_T = 2\pi fL_1 + 2\pi fL_2 + 2\pi fL_3 + \ldots\ldots + 2\pi fL_N$ or
$X_{LT} = X_{L1} + X_{L2} + X_{L3} + \ldots + X_{LN}$.
The total inductive reactance is equal to the sum of the individual reactances.

Similarly consider inductive reactances in parallel. The following equation is for inductors in parallel:

$$\frac{1}{L_T} = \frac{1}{L_1} + \frac{1}{L_2} + \frac{1}{L_3} + \ldots + \frac{1}{L_N}$$

Therefore:
$$\frac{1}{2\pi fL_T} = \frac{1}{2\pi fL_1} + \frac{1}{2\pi fL_2} + \frac{1}{2\pi fL_3} + \ldots + \frac{1}{2\pi fL_N}$$

$$\frac{1}{X_{LT}} = \frac{1}{X_{L1}} + \frac{1}{X_{L2}} + \frac{1}{X_{L3}} + \ldots + \frac{1}{X_{LN}}$$

The formulas for inductive reactances and resistances are therefore the same.

Example 9-7

In Fig. 9-14, $e = 12 \sin\left(5000t + \frac{\pi}{3}\right)$ volts and L = 850 mH. Calculate the values of the effective current and the reactive power. Write down the trigonometrical expression for the instantaneous current. If the frequency is doubled, what is the new value of the effective current?

Solution

Effective applied voltage, $E_{rms} = \dfrac{12}{\sqrt{2}} = 8.49\,V$.

Angular velocity, $\omega = 5000$ radians per second.
Inductive reactance, $X_L = \omega L = 5000 \times 850 \times 10^{-3} = 3250\,\Omega$.

Effective current, $I_{rms} = \dfrac{E_{rms}}{X_L} = \dfrac{8.49\,V}{3250\,\Omega} = 2.61\,mA$.

Average reactive power $= E_{rms} \times I_{rms} = 22.16\,mVArs$. Since e

leads i by 90° or $\dfrac{\pi}{2}$ radians, the instantaneous current, i = 2.61 ×

$1.414 \, Sin \, (5000t + \dfrac{\pi}{3} - \dfrac{\pi}{2}\,) = 3.69 \, Sin \, (5000 - \dfrac{\pi}{6}\,)\, mA.$

If the frequency is doubled, the inductive reactance is doubled and the effective current is halved. The new effective current is therefore

$$\dfrac{2.61}{2} = 1.305 \, mA.$$

Example 9-8

A 750 kHz sine wave voltage whose effective value is 5 V, is applied across an inductor. If the effective current is 1.75 mA, calculate the value of the inductor.

Solution

Inductive reactance, $X_L = \dfrac{E_{rms}}{I_{rms}} = \dfrac{5\,V}{1.75\,mA} = 2857\Omega$.

Inductance, $L = \dfrac{X_L}{2\pi f} = \dfrac{2857}{2 \times \pi \times 750 \times 10^3}$ H $= 606\,mH.$

Example 9-9

Three inductors whose reactances are 150Ω, 275Ω and 310Ω, are connected in series across a 110 V (rms), 60 Hz source.

Calculate the effective current and the individual voltages across the inductors. What is the value of the total inductance?

Solution

Total inductive reactance $X_{LT} = 150 + 275 + 310 = 735\,\Omega$.

Effective current $= \dfrac{110\,V}{735\,\Omega} = 0.1496\,A$.

The individual voltages across the inductors are $150\,\Omega \times 0.1496\,A = 22.45\,V$, $275\,\Omega \times 0.1496\,A = 41.16\,V$ and $310\,\Omega \times 0.1496\,A = 46.39\,V$.

Total inductance $= \dfrac{X_{LT}}{2\pi f} = \dfrac{735}{2 \times \pi \times 60} = 1.95\,H$.

CAPACITANCE IN AN AC CIRCUIT

Let an AC voltage $e = E_{max} \, \mathrm{Sin}\,\omega t$ be applied to a capacitor C (Fig. 9-15). Then $v_c = -\dfrac{q}{C}$ and $e + v_c = 0$. Therefore: $e = \dfrac{q}{C}$

and $q = Ce = CE_{max}\,\mathrm{Sin}\,\omega t$
Differentiating with respect to t:

$$i = \frac{dq}{dt} = \omega\, CE_{max}\,\mathrm{Cos}\,\omega t = \omega C E_{max}\,\mathrm{Sin}\left(\omega t + \frac{\pi}{2}\right) = I_{max}\,\mathrm{Sin}$$

$\left(\omega t + \dfrac{\pi}{2}\right)$ e therefore lags i by $\dfrac{\pi}{2}$ radians or $90°$ so that the

phase angle, ϕ, is $-90°$. Alternatively, i leads e by $90°$ which may be remembered by the word "iCe"; this may be combined with "eLi" to form "eLi, the iCe man".

Since $\omega c\, E_{max} = I_{max}$, $\dfrac{E_{max}}{I_{max}} = \dfrac{E_{rms}}{I_{rms}} = \dfrac{1}{\omega C} = \dfrac{1}{2\pi f C} = X_c\,\Omega$.

Then: $I_{rms} = 2\pi f C E_{rms}$

and $C = \dfrac{1}{2\pi f X_c}$, $f = \dfrac{1}{2\pi C X_c}$

The quantity X_c is called the capacitive reactance which measures the opposition to alternating current offered by a capacitor. Notice that X_c is inversely proportional to f so that the graphs of I_{rms} and X_c against frequency are as shown in Fig. 9-15.

Since X_L is directly proportional to frequency while X_C is inversely proportional to frequency, there must be, for particular values of L and C, a certain frequency for which $X_L = X_C$.

The instantaneous power is:

$$p = ei = E_{max} I_{max} \sin \omega t \sin (\omega t + \frac{\pi}{2})$$
$$= \frac{E_{max} I_{max}}{2} \sin 2 \omega t$$

The average power over the complete cycle is zero. During two quarters of the cycle, energy is taken from the source in charging the capacitor but during the remainder of the cycle, the capacitor discharges and the energy is returned to the source.

Now consider capacitive reactances in series. The following equation is for N capacitors in series:

$$\frac{1}{C_T} = \frac{1}{C_1} + \frac{1}{C_2} + \frac{1}{C_3} + \cdots + \frac{1}{C_N}$$

Then:

$$\frac{1}{2 \pi f C_T} = \frac{1}{2 \pi f C_1} + \frac{1}{2 \pi f C_2} + \frac{1}{2 \pi f C_3} + \cdots + \frac{1}{2 \pi f C_N}$$

or $X_{CT} = X_{C1} + X_{C2} + X_{C3} + \cdots + X_{CN}$.

Although the reciprocal formula is required for capacitances in series, the total capacitive reactance is merely the sum of the individual reactances.

Now consider capacitive reactances in parallel. The following equation is for N capacitors in parallel:
$$C_T = C_1 + C_2 + C_3 + \cdots + C_N.$$

Then:
$$\frac{2 \pi f C_T}{1} = \frac{2 \pi f C_1}{1} + \frac{2 \pi f C_2}{1} + \frac{2 \pi f C_3}{1} + \cdots + \frac{2 \pi f C_N}{1}$$

or
$$\frac{1}{X_{CT}} = \frac{1}{X_{C1}} + \frac{1}{X_{C2}} + \frac{1}{X_{C3}} + \cdots + \frac{1}{X_{CN}}$$

Although capacitances in parallel are simply added, the reciprocal or repeated product-over-sum formula is required to find the total capacitive reactance.

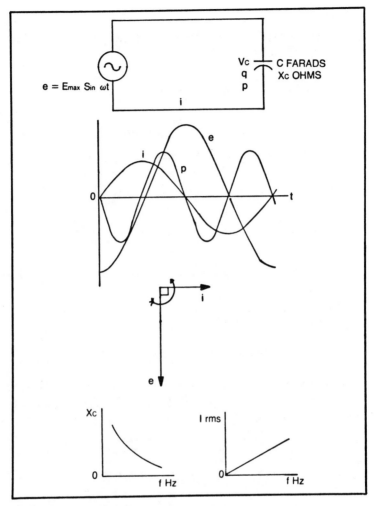

Fig. 9-15. Capacitance in ac circuits.

Example 9-10

In Fig. 9-15, $e = 17 \sin(2\pi \times 3.25 \times 10^3 t - \dfrac{\pi}{12})$ volts and C

$= 0.15\ \mu F$. Calculate the value of the rms current and write down the trigonometrical expression for the instantaneous current. IF a further $0.33\ \mu f$ capacitor is added in parallel with C and the frequency is then tripled, what is the new value of the effective current?

173

Solution

Frequency, $f = 3.25 \times 10^3$ Hz and the effective applied voltage is $17/\sqrt{2} = 12.02$ V.

Capacitive reactance, $X_C = \dfrac{1}{2\pi fC}$

$$= \frac{1}{2 \times \pi \times 3.25 \times 10^3 \times 0.15 \times 10^{-6}}$$

$$= 326\,\Omega.$$

Effective current, $I_{rms} = \dfrac{E_{rms}}{X_C} = \dfrac{12.02V}{326\Omega} = 36.83\,mA$,

i leads e by $\dfrac{\pi}{2}$

radians and therefore the instantaneous current is $i = \dfrac{17}{326}$ Sin (2π

$\times 3.25 \times 10^3 t - \dfrac{\pi}{12} + \dfrac{\pi}{2}$) A

$$= 52.1 \text{ Sin } (2\pi \times 3.25 \times 10^3 t - \frac{5\pi}{12}) \text{ mA}.$$

When the 0.33 μF capacitor is added in parallel, the total capacitance, C_T, is $0.33 + 0.15 = 0.48$ μF. The frequency is changed to $3 \times 3.25 \times 10^3 = 9.75 \times 10^3$ Hz so that $X_{CT} =$

$$\frac{1}{2\pi \times 9.75 \times 10^3 \times 0.48 \times 10^{-6}} = 34\,\Omega.$$

The new effective current is $12.02/34 = 0.354\,A$.

Example 9-11

Three capacitors of values of 150 pFs, 220 pFs and 180 pFs are connected in series across a 12 V rms, 1 MHz source. Calculate the individual voltages across the capacitors.

Solution

The individual capacitive reactances are

$$X_{C1} = \frac{1}{2\pi fC_1} = \frac{1}{2 \times \pi \times 1 \times 10^6 \times 150 \times 10^{-12}} = 1061\,\Omega.$$

$$X_{C2} = X_{C1} \times \frac{C_1}{C_2} \quad 1061 \times \frac{150}{220} = 723\,\Omega.$$

$$X_{C3} = 1061 \times \frac{150}{180} = 884\,\Omega.$$

Total capacitive reactance, $X_{CT} = 1061 + 723 + 884 = 2668\,\Omega$

Alternatively:

$$\frac{1}{C_T} = \frac{1}{150} + \frac{1}{220} + \frac{1}{180}$$

$$= 0.00667 + 0.00455 + 0.00556$$
$$= 0.016767$$

and $X_{CT} = \dfrac{1}{2\pi f C_T} = \dfrac{0.016767 \times 10^{12}}{2 \times \pi \times 1 \times 10^6} = 2668\,\Omega.$

Using the voltage division rule:

$$V_{C1} = 12\,V \times \frac{1061\,\Omega}{2668\,\Omega} = 4.77\,V.$$

$$V_{C2} = 12\,V \times \frac{723\,\Omega}{2668\,\Omega} = 3.25\,V.$$

$$V_{C3} = 12\,V \times \frac{884\,\Omega}{2668\,\Omega} = 3.98\,V.$$

Check:
$$E = V_{C1} + V_{C2} + V_{C3} = 12.00\,V.$$

THE GENERAL AC CIRCUIT

The general AC circuit will contain resistance, inductive reactance and capacitive reactance. The total opposition to the flow of alternating current is therefore a resistance/reactance combination which is called the *impedance*, z.

Z is a phasor which is defined as the ratio of the source voltage phasor to the source current phasor. The magnitude of z is measured in ohms and is given by:

$$Z = \frac{E_{rms}}{I_{rms}} = \frac{E_{max}}{I_{max}}$$

where E_{rms} and I_{rms} are the effective values of the source voltage and the source current.

The direction of z is measured by the angle, ϕ, between the z phasor and the horizontal line. The value of this angle normally extends from $-\ 90°$ (capacitive reactance only) through $0°$ (resistance only) to $+\ 90°$ (inductive reactance only).

Since the source voltage and the source current are $\phi°$ out of phase, the current may be represented by $i = I_{max}$ Sinωt and the voltage will then be $e = E_{max}$ Sin $(\omega t \pm \phi)$. The instantaneous power is: $p = ei = E_{max} I_{max}$ Sin ωt Sin $(\omega t \pm \phi)$.

$$= \frac{E_{max} I_{max}}{2} (\text{Cos } \phi - \text{Cos } (2\omega t \pm \phi)).$$

The mean value of Cos $(2\omega t \pm \phi)$ is zero and therefore the average power over the complete cycle is

$$\frac{E_{max} I_{max} \text{Cos } \phi}{2} = E_{rms} \times I_{rms} \text{ Cos } \phi \text{ watts.}$$

This is the true power dissipated in the resistive part of the impedance. By contrast, the product $E_{rms} \times I_{rms}$ is called the apparent power at the source and is measured in voltamperes (VA). The cosine of the phase angle, Cos ϕ, is called the pwoer factor which is defined by: Power factor, Cos $\phi =$

$$\frac{\text{True Power, Watts}}{\text{Apparent Power, Volt-amperes}},$$

or True Power $= E_{rms} \times I_{ms} \times$ power factor.

If an AC current is inductive, the phase angle is positive and the power factor is lagging (source current lags source voltage). With a capacitive circuit the phase angle is negative and the power factor is leading. The value of the power factor extends from zero (reactance only) to 1 (resistance only).

Example 9-12

An alternating source has an instantaneous voltage of $e = 28$ Sin $(\omega t + \pi 4)$ volts and is connected across an AC circuit. If the instantaneous current drawn from the source is

$$5.7 \text{ Sin} (\omega t - \frac{\pi}{12})$$

mA, calculate the values of the impedance, the phase angle and the power factor. What are the values of the true power and the apparent power?

Solution

The impedance, $Z = \dfrac{E_{rms}}{I_{rms}} = \dfrac{28 \, V}{5.7 \, mA} = 4.91 \, k\Omega$.

e leads i by $\phi = \dfrac{\pi}{4} - (-\dfrac{\pi}{12}) = \dfrac{+\pi}{3}$ *radians*.

The power factor is Inv. Cos. $\dfrac{+\pi}{3} = 0.5 \, lagging$.

Apparent power $= 28 \, V \times 5.7 \, mA = 159.6 \, mVA$.

True power $=$ Apparent power $\times$ power factor
$$= 159.6 \times 0.5$$
$$= 79.8 \, mW.$$

CHAPTER SUMMARY

□ *Sinewave Relationships*

Period, T (seconds) $= \dfrac{1}{\text{Frequency, f, (Hz)}}$.

Frequency, f(Hz) $= \dfrac{1}{\text{Period, T(seconds)}}$.

Generated frequency, $f = \dfrac{N \, (rpm) \times p \, (\text{number of pairs of poles})}{60}$ Hz.

Angular velocity, $\omega = 2\pi f = \dfrac{2\pi}{T}$ radians per second.

One cycle $= 360° = 2\pi$ radians.

ϕ (degrees) $= \dfrac{180}{\pi} \times \phi$ (radians)
$$= 57.296 \times \phi \text{ (radians)}.$$

ϕ (radians) $= \dfrac{\pi}{180} \times \phi$ (degrees) $= 0.01745 \times \phi$ (degrees).

□ *Sinewave equations*

Instantaneous voltage, $e = E_{max} \, Sin \, \omega t = E_{max} \, Sin \, 2\pi \, ft$

$$= E_{max} \sin \phi.$$

$$E_{max} \text{ (peak value)} = \frac{E_{p-p}}{2} = E_{rms} \times \sqrt{2} = E_{rms} \times 1.414$$

$$= E_{av} \times \frac{\pi}{2}$$

$$= E_{av}$$

$$E_{rms} \text{ (effective value)} = E_{p-p}/(2\sqrt{2}) = E_{p-p} \times 0.3535$$
$$= E_{max}/\sqrt{2} = E_{max} \times 0.707$$
$$= E_{av} \times \pi/(2\sqrt{2}) = E_{av} \times 1.11.$$

$$E_{p-p} \text{ (peak-to-peak value)} = 2 E_{max} = 2\sqrt{2} E_{rms} = 2.828 E_{rms}$$
$$= E_{av} \times \pi \times E_{av} \times 3.1416.$$

F_{av} (average over an alternation)

$$= E_{p-p}/\pi = E_{p-p} \times 0.318$$
$$= E_{rms} \times 2\sqrt{2}/\pi = E_{rms} \times 0.9$$
$$= E_{max} \times 2/\pi = E_{max} \times 0.637.$$

□ *Phase Difference*

$e_1 = E_{1max} \sin \omega t$ lags $e_2 = E_{2max} \sin(\omega t + \phi)$
by ϕ radians (e_2 leads e_1 by ϕ radians).

$$e_1 + e_2 = \sqrt{E_{1max}^2 + E_{2max}^2 + 2E_{1max} E_{2max} \cos \phi} \, \sin(\omega t + \phi)$$

where ϕ, = Inv. Tan. $\left(\dfrac{E_{2max} \sin \phi}{E_{1max} + E_{2max} \cos \phi} \right)$

$$e_1 - e_2 = \sqrt{E_{1max}^2 + E_{2max}^2 - 2E_{1max} E_{2max} \cos\phi} \, \sin(\omega t + \phi_2)$$

where ϕ_2 = Inv. Tan. $\left(\dfrac{E_{2max} \sin\phi}{E_{2max} \cos\phi - E_{1max}} \right)$

□ *Resistance in AC*

e and i are in phase. $\phi = 0°$.

$$\frac{E_{rms}}{I_{rms}} = R\Omega.$$

Average power $= E_{rms} \times I_{rms} = \dfrac{E_{rms}^2}{R} = I_{rms}^2 \times R$ watts.

□ *Inductance in AC*

e leads i by $\dfrac{\pi}{2}$ radians. $\phi = +90°$ $\dfrac{E_{rms}}{I_{rms}} = 2\pi fL = \omega L = X_L \, \Omega.$

Average power = 0. Reactive power = $E_{rms} \times I_{rms}$ VArs.
Inductive reactances in series:

$$X_{LT} = X_{L1} + X_{L2} + X_{L3} + \text{-----} + X_{LN}.$$

Inductive reactances in parallel:

$$\frac{1}{X_{LT}} = \frac{1}{X_{L1}} + \frac{1}{X_{L2}} + \frac{1}{X_{L3}} + \text{----} + \frac{1}{X_{LN}}$$

□ *Capacitance in AC*

i leads e by $\dfrac{\pi}{2}$ radians. $\phi = -90°$.

$$\frac{E_{rms}}{I_{rms}} = \frac{1}{2\pi fC} = \frac{1}{\omega C} = X_c \, \Omega.$$

Average power = 0. Reactive power = $E_{rms} \times I_{rms}$ VArs.
Capacitive reactances in series:

$$X_{CT} = X_{C1} + X_{C2} + X_{C3} + \text{------} + X_{CN}.$$

Capacitive reactances in parallel:

$$\frac{1}{X_{CT}} = \frac{1}{X_{C1}} + \frac{1}{X_{C2}} + \frac{1}{X_{C3}} + \text{------} + \frac{1}{X_{CN}}.$$

□ *General AC circuit*

Magnitude of impedance, $Z = E_{rms}/I_{rms}$ ohms.

Apparent power = $E_{rms} \times I_{rms}$ VA.

$$\text{Power Factor} = \frac{\text{True Power}}{\text{Apparent Power}} = \text{Cos} \, \phi.$$

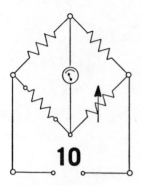

Alternating Current Circuits

10

Referring to Fig. 10-1, voltage V_R is in phase with i, which lags V_L by 90°. The phasor sum of V_R and V_L is the supply voltage, e, which leads i (the current flowing through both R and L) by the phase angle ϕ.

SINEWAVE INPUT TO L AND R IN SERIES

The total opposition to current flow is measured by the impedance, z, which is defined as the ratio of e to i and is the phasor sum of R and X_L. The magnitude of z is:

$$Z = E/I = \sqrt{(R^2 + X_L^2)} = \sqrt{(R^2 + 4\pi^2 f^2 L^2)}$$

and the rms voltages are:

$$V_R = IR$$

$$V_L = IX_L$$

$$E = \sqrt{(V_R^2 + V_L^2)}$$

It follows that:

$$R = \sqrt{(Z^2 - X_L^2)}$$

$$X_L = \sqrt{Z^2 - R^2)}$$

$$V_R = \sqrt{(E^2 - V_L^2)}$$

and $V_L = \sqrt{(E^2 - V_R^2)}$

The instantaneous power, p, is equal to e × i; its waveform is a second harmonic curve as shown in Fig. 10-1.

Only the resistance in the circuit dissipates power; thus the apparent power in the circuit is

$$EI = \frac{E^2}{Z} = I^2 Z$$

and is measured in volt-amperes (VA). The idle, wattless or reactive power associated with the inductor is:

$$I^2 X_L = \frac{V_L^2}{X_L} = IV_L$$

and is measured in voltamperes reactive (VArs). The true power in watts is:

$$I^2 R = \frac{V_R^2}{R} = IV_R$$

Then: Apparent Power = $\sqrt{\text{True Power}^2 + \text{Reactive Power}^2}$. This relationship may be derived from the power phasor diagram in Fig. 10-1.

The power factor is defined as the ratio of the true power to the apparent power. Therefore, the power factor, which has no units, is:

$$\frac{\text{True Power}}{\text{Apparent Power}} = \frac{I^2 R}{I^2 Z} = \frac{R}{Z} = \cos \phi.$$

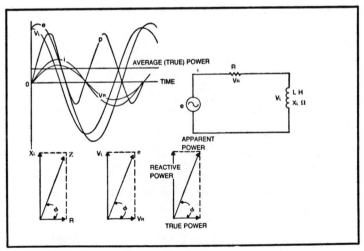

Fig. 10-1. L and R in series; X_L is greater than R.

Also:

$$\sin \phi = \frac{X_L}{Z} = \frac{\text{Reactive Power}}{\text{Apparent Power}}$$

and

$$\tan \phi = \frac{X_L}{R} = \frac{\text{Reactive Power}}{\text{True Power}}$$

The true power in watts is EI × the power factor. Note that a wattmeter used for AC measurements automatically takes into account the power factor and indicates the true power. It follows that:

$$I = \frac{\text{true power}}{E \times \text{power factor}}$$

and

$$E = \frac{\text{true power}}{I \times \text{power factor}}$$

If $R = X_L$, the phase angle of the supply voltage relative to the supply current is $+45°$ and the power factor is 0.707. With ϕ a positive angle, the power factor is said to be lagging. For an entirely resistive circuit, $\phi = 0°$ and the power factor is unity; for an entirely reactive circuit, $\phi = \pm 90°$ and the power factor is zero. Under no circumstances can the power factor of a circuit exceed 1.

Example 10-1

In Fig. 10-1, $L = 150\mu H$, $R = 680\,\Omega$ and the supply voltage is 1 V, 800 kHz., Calculate the values of Z, V_L, V_R, ϕ, true power, apparent power and power factor.

Solution

Inductive reactance, $X_L = 2\pi fL = 2 \times \pi \times 800 \times 10^3$
$\times 150 \times 10^{-6}$
$= 754\,\Omega.$

Impedance, $Z = \sqrt{R^2 + X_L^2} = \sqrt{680^2 + 754^2} = 1015\,\Omega.$

Current, $I = \dfrac{E}{Z} = \dfrac{1V}{1015\,\Omega} = 0.985\,mA.$

$V_L = IX_L = 0.985\,mA \times 754\,\Omega = 0.743\,V.$
$V_R = IR = 0.985\,mA \times 680\,\Omega = 0.670\,V.$

Voltage check: $E = \sqrt{V_R^2 + V_L^2} = \sqrt{0.670^2 + 0.743^2} = 100\,V.$

Apparent Power $= E \times I = V \times 0.985\,mA = 0.985\,mVA.$

Reactive Power $= V_L \times I = 0.743\,V \times 0.985\,mA = 0.732$ mVAr.

True Power $= V_R \times I = 0.670\,V \times 0.985\,mA = 0.660\,mW.$
Power check:

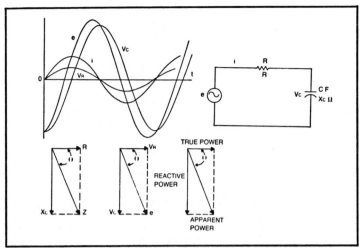

Fig. 10-2. C and R in series; X_C is greater than R.

$$\text{Apparent Power} = \sqrt{(\text{True Power})^2 + (\text{Reactive Power})^2}$$
$$= \sqrt{0.660^2 + 0.732^2}$$
$$= 0.986\, mVA.$$

The slight difference between 0.985 mVA and 0.986 mVA is due to rounding off.

$$\text{Power factor} = \frac{\text{True Power}}{\text{Apparent Power}} = \frac{0.660\,\text{mW}}{0.985\,\text{mVA}} = 0.67,$$

lagging.

Phase angle, $\phi = \text{Inv. Cos. } 0.67 = +47.9°$

SINEWAVE INPUT TO C AND R IN SERIES

Referring to Fig. 10-2, the equations for C and R in series may be derived from those given earlier by substituting:

$$X_C = \frac{1}{2\pi fC} = \frac{0.159}{fC}\ \ \Omega.$$

for X_L. The only differences are:

1. V_C lags i by 90°.
2. e lags i by the phase angle, ϕ, which now negative.
3. The power factor is considered to be leading.

Example 10-2

In Fig. 10-2, C = 0.068 μF, R = 820Ω and the supply voltage is 15 V, 4.5 kHz. Calculate the values of Z, V_C, V_R, ϕ, true power, apparent power and power factor.

Solution

Capacitive reactance, $X_C = \frac{1}{2\pi fC}$

183

$$= \frac{1}{2 \times \pi \times 4.5 \times 10^3 \times 0.068 \times 10^{-6}}$$

$$= 520\Omega.$$

Impedance, $Z = \sqrt{R^2 + X_C^2} = \sqrt{820^2 + 520^2} = 971\Omega.$

Current, $I = \dfrac{E}{Z} = \dfrac{15\text{ V}}{971\Omega} = 15.45$ mA.

$\quad V_C = IX_C = 15.45$ mA $\times 520\Omega = 8.03$ V.

$\quad V_R = IR = 15.45$ mA $\times 820\Omega = 12.67$ B.

Voltage check: $E = \sqrt{V_C^2 + V_R^2} = \sqrt{8.03^2 + 12.67^2} = 15.00$

V.

Apparent Power $= E \times I = 15$ V $\times 13.45$ mA $= 231.75$ mVA.

Reactive Power $= V_C \times I = 8.03 \times 15.45$ mA $= 124.06$ mVAr.

True Power $= V_R \times I = 12.67$ V $\times 15.45$ mA $= 195.75$ mW.

Power check:

Apparent Power $= \sqrt{(\text{True Power})^2 + (\text{Reactive Power})^2}$
$$= \sqrt{195.75^2 + 124.06^2}$$
$$= 231.75 \text{ mVA}.$$

Power Factor $=$

$\dfrac{\text{True Power}}{\text{Apparent Power}} = \dfrac{195.75}{231.75} = 0.845$, leading

Phase angle, $\phi =$ Inv. Cos. $0.845 = -32.4°$.

SINEWAVE INPUT TO L AND C IN SERIES

Referring to Fig. 10-3, V_L leads i by 90°. Therefore v_L and v_C are 180° out of phase and when added together to produce their phasor sum, e, their magnitudes, V_L and V_C, will tend to cancel. If X_L is greater than X_C, the total impedance, $Z = X_L - X_C$, the source voltage, $E = V_L - V_C$ and e leads i by 90° ($\phi = +90°$); the circuit behaves inductively. If X_L is less than X_C, $Z = X_C - X_L$, $E = V_C - V_{ol}$, and e lags i by 90° ($\phi = -90°$); the circuit behaves capacitively. In both cases the circuit power factor is zero (it is assumed that there are no losses associated with either the inductor or the capacitor). If $X_L = X_C$, $V_L = V_C$, $Z = 0$, I is theoretically infinite, and the series combination of L and C therefore behaves as a short circuit ($\phi = 0°$). For particular values of L and C, this situation will occur at a frequency

$$f = \frac{1}{2\pi\sqrt{LC}} = \frac{0.159}{\sqrt{LC}}$$

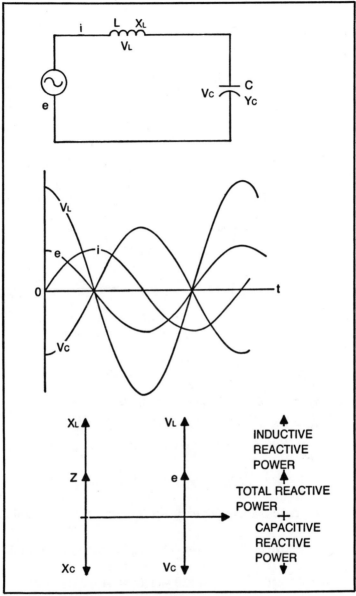

Fig. 10-3. L and C in series; X_L is greater than X_C.

where f is in Hz, L in henrys, C in farads

Then: $L = \dfrac{1}{4\pi^2 f^2 C} = \dfrac{0.0253}{f^2 C}$ H

and
$$C = \frac{1}{4\pi^2 f^2 L} = \frac{0.0253}{f^2 L} \text{ F}$$

Example 10-3

In Fig. 10-3, L = 125 μH, C = 275 pFs and the source voltage is 100 mV, 750 kHz. Calculate the values of Z, I, V_L, V_C and the total reactive power.

Solution

Inductive reactance, $X_L = 2\pi fL = 2 \times \pi \times 750 \times 10^3$
$$\times 125 \times 10^{-6}$$
$$= 589\Omega.$$

Capacitance reactance,

$$X_C = \frac{1}{2\pi fC}$$
$$= \frac{1}{2 \times \pi \times 750 \times 10^3 \times 275 \times 10^{-12}}$$
$$= 772\Omega.$$

Impedance, $Z = X_C - X_L = 772 - 589 = 183\Omega$ and is capacitive.

Current, $I = \dfrac{E}{Z} = \dfrac{100 \text{ mV}}{183\Omega} = 0.546$ mA.

$V_L = I \times X_L = 0.546$ mA $\times 589\Omega = 322$ mV.

$V_C = I \times X_C = 0.546$ mA $\times 772\Omega = 422$ mV.

Voltage check: Source Voltage, $E = V_C - V_L = 422 - 322 = 100$ mV.

Notice that V_L and V_C are each greater than the source voltage.

Inductive reactive power = $I \times V_L = 0.546$ mA $\times 322$ mV
$$= 176 \text{ } \mu\text{VArs.}$$

Capacitive reactive power = $I \times V_C = 0.546$ mA $\times 422$ mV
$$= 230 \text{ } \mu\text{VArs.}$$

Total reactive power = $230 - 176 = 54$ μVArs.

Power check: Total reactive power = $E \times I = 100$ mV $\times 0.546$ mA
$$= 54.6 \text{ } \mu\text{VArs.}$$

The power factor is zero, leading, and the phase angle is $-90°$.

SINEWAVE INPUT TO L, C AND R IN SERIES

Referring to Fig. 10-4, V_R is in phase with i, V_L leads i by 90° and V_C lags i by 90°. Assuming that X_L is greater than X_C, the phasor sum, V_X, of V_L and V_{tc} is in phase with V_L while e (which is the phasor sum of V_L, V_C, and V_R) leads i by the phase angle ϕ. The circuit, therefore, behaves inductively and is positive. If X_C is greater than X_L, the phasor sum of V_L and V_C is in phase with V_C, e lags i, the circuit is capacitive and ϕ is negative.

Note that the special case of $X_L = X_C$ is concerned with the phenomenon of resonance, which is covered in chapter 11.

The magnitude of the impedance is $Z = \sqrt{(R^2 + X^2)}$ where X is the net reactance which is the difference in value between X_L and X_C.

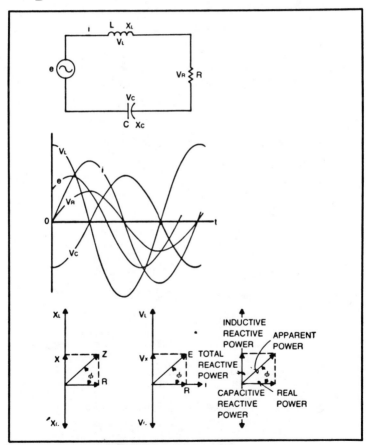

Fig. 10-4. L, C and R in series; X_L is greater than X_C.

The supply voltage is $E = \sqrt{(V_R{}^2 + V_X{}^2)}$, where V_X is the difference in value between V_L and V_C. The current is: $I = E/Z$ and $V_R = IR$, $V_L = IX_L$, $V_C = IX_C$.

Note that although V_R cannot exceed the supply voltage, E, either V_L or V_C (or both V_L and V_C) can be greater than E.

Apparent power = $EI = I^2Z = E^2/Z$ volt-amperes.

True power in watts = $I^2R = V_R{}^2/R = IV_R = EI \times$ power factor.

$$\text{Power factor} = \frac{\text{True Power}}{\text{Apparent Power}} = \text{Cos } \phi = R/Z$$

and $\text{Sin } \phi = \dfrac{X}{R}$, $\text{Tan } \phi = \dfrac{X}{R}$

If R = X, the phase angle is 45°.

Example 10-4

In Fig. 10-4, L = 2.5 H, C = 5 μF and R = 350 Ω. If the source voltage is 110 V, 60 Hz, calculate the values of the total impedance, V_L, V_C, V_R, the power factor and the phase angle.

Solution

Inductive reactance, $X_L = 2\pi fL = 2 \times \pi \times 60 \times 2.5$
$$= 942.5 \ \Omega$$

Capacitive reactance, $X_C = \dfrac{1}{2\pi fC}$

$$= \frac{1}{2 \times \pi \times 60 \times 5 \times 10^{-6}}$$

$$= 530.5 \ \Omega.$$

Total impedance, $Z = \sqrt{R^2 + (X_L - X_C)^2}$

$$= \sqrt{350^2 + (942.5 - 530.5)^2}$$

$$= 541 \ \Omega.$$

Since X_L is greater than X_C, the impedance is inductive.

Current, $I = \dfrac{E}{Z} = \dfrac{110 \text{ V}}{541 \Omega} = 0.204$ A.

$V_L = I \times X_L = 0.204 \text{ A} \times 942.5 \ \Omega = 191.8 \ V.$

$V_C = I \times X_C = 0.204 \text{ A} \times 530.5 \ \Omega = 108.2 \ V.$

$V_R = I \times R = 0.204 \text{ A} \times 350 \ \Omega = 71.4 \ V.$

Voltage check:

Source voltage, $E = \sqrt{71.4^2 + (191.8 - 108.2)^2}$

$$= \sqrt{71.4 + 83.6^2} = 110 \text{ V.}$$

True Power $= I \times V_R = 0.204 \text{ A} \times 71.4 \text{ V} = 14.6 \text{ W.}$

Reactive Power $= I \times (V_L - V_C) = 0.204 \text{ A} \times 83.6 \text{ V}$
$$= 17.1 \text{ VAr.}$$

Apparent Power $= I \times E = 0.204 \text{ A} \times 110 \text{ V} = 22.4 \text{ VA.}$

Power check: Apparent Power $= \sqrt{14.6^2 + 17.1^2} = 22.5 \text{ VA.}$

Power Factor $= \dfrac{\text{True Power}}{\text{Apparent Power}} = \dfrac{14.6}{22.5} = 0.65$, lagging.

Phase angle, $\phi = $ Inv. Cos 0.65 $= +49.5°.$

SINEWAVE INPUT TO L AND R IN PARALLEL

Referring to Fig. 10-5, current i_R is in phase with e while i_L lags e by 90°. The phasor sum of i_L and i_R is the supply current, i_s, which lags e by the phase angle, ϕ.

The magnitude of the total impedance is given by the product-over-sum formula:

$$Z = \frac{RX_L}{\sqrt{R^2 + X_L^2}}$$

Note that Z is smaller in value than either R or X_L.

The supply current is:

$$I_s = \frac{E}{Z} = \sqrt{(I_R^2 + I_L^2)}.$$

The lagging power factor is:

$$\text{cosine } \phi = \frac{Z}{R} (not \frac{R}{Z}).$$

The apparent power is:

$$EI_s \text{ volt-amperes.}$$

The true power is:

$$EI_s \times \text{power factor} = I_R^2 R = \frac{E^2}{R} \text{ watts}$$

$$\text{Also Sin } \phi = \frac{Z}{X_L} \text{ and Tan } \phi = \frac{R}{X}.$$

For parallel circuits it may be more convenient to work with conductance, G, susceptance, B, and admittance, Y, which are, respectively, the reciprocals of resistance, reactance, and impedance, and are all measured in siemens. Therefore the conductance is $G = \frac{1}{R}$ (as in DC circuits), the inductive susceptance, is $B_L = \frac{1}{X_L}$ and the admittance, which is the phasor sum of the conductance and the susceptance, is $Y = \frac{1}{Z}$. In terms of their magnitudes:

$$\text{Sin} \phi = \frac{B_L}{Y}, \text{Cos} \phi = \frac{G}{Y}, \text{Tan} \phi = \frac{B_L}{G}$$

$$Y = \sqrt{G^2 + B_L^2} = \sqrt{\frac{1}{R^2} + \frac{1}{X_L^2}}$$

$$= \frac{\sqrt{R^2 + X_L^2}}{RX_L} = \frac{1}{Z}$$

Example 10-5

In Fig. 10-5, L = 550 mH and R = 6.8 k Ω. If the source voltage is 8 V, 2.5 kHz, calculate the values of Z, I_R, I_L, I_S, true power, apparent power, power factor and phase angle.

Solution

Inductive reactance, $X_L = 2\pi f L$

$$= 2 \times \pi \times 2.5 \times 10^3 \times 550 \times 10^{-3} \, \Omega$$

$$= 8.64 \text{ k} \, \Omega.$$

To calculate the total impedance:

Method 1.

Total impedance, $Z = \frac{R \times X_L}{\sqrt{R^2 + X_L^2}} = \frac{6.8 \times 8.64}{\sqrt{6.8^2 + 8.64^2}}$

$$= 5.34 \, k\Omega.$$

Method 2

$$I_R = \frac{8V}{6.8 \text{ k } \Omega} = 1.18 \text{ mA}.$$

$$I_L = \frac{8 V}{8.64 \text{ k } \Omega} = 0.926 \text{ mA}.$$

$$I_S = \sqrt{1.18^2 + 0.926^2} = 1.5 \text{ mA}.$$

Total impedance, $Z = \dfrac{E}{I_S} = \dfrac{8 V}{1.5 \text{ mA}} = 5.33 \text{ k } \Omega.$

If no source voltage is given, method 2 may be used by assuming any convenient value for E.

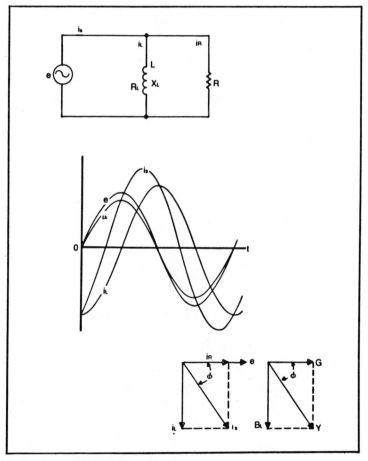

Fig. 10-5. L and R in parallel.

Method 3

Inductive susceptance, $B_L = \dfrac{1}{8.64 \text{ k}\Omega} = 1.16 \times 10^{-4}$ siemens

Conductance, $G = \dfrac{1}{6.8 \text{ k}\Omega} = 1.47 \times 10^{-4}$ siemens,

Admittance, $\begin{aligned} Y &= \sqrt{G^2 + B_L^{\,2}} \\ &= \sqrt{(1.47 \times 10^{-4})^2 + (1.16 \times 10^{-4})^2} \\ &= 1.87 \times 10^{-4} \text{ siemens.} \end{aligned}$

Impedance, $Z = \dfrac{1}{Y} = \dfrac{1}{1.87 \times 10^{-4}} = 5.34 \text{ k}\,\Omega.$

True Power $= \dfrac{E^2}{R} = \dfrac{(8 \text{ V})^2}{6.8 \text{ k}\Omega} = 9.41 \text{ mW.}$

Note that the true power is constant and independent of frequency.

Reactive power $= \dfrac{E^2}{X_L} = \dfrac{(8 \text{ V})^2}{8.64 \text{ k}\Omega} = 7.41 \text{ mVArs.}$

Apparent power $= \dfrac{E^2}{Z} = \dfrac{(8 \text{ V})^2}{5.34 \text{ k}\Omega} = 11.99 \text{ mVA.}$

Power check: Apparent power $= \sqrt{(\text{True Power})^2 + (\text{Reactive Power})^2}$
$$= \sqrt{9.41^2 + 7.41^2}$$
$$= 11.98 \text{ mVA.}$$

Power factor $= \dfrac{\text{True Power}}{\text{Apparent Power}} = \dfrac{9.41}{11.99} = 0.8.$

$$= \dfrac{Z}{R} = \dfrac{5.34}{6.8} = 0.8, \text{ lagging}$$

Phase angle, $\phi = $ Inv. Cos. $0.8 = +38°.$

SINEWAVE INPUT TO C AND R IN PARALLEL

The equations for C and R in parallel may be derived from those given earlier, by substituting X_C for X_L, I_C for I_L and B_C for B_L. Referring to Fig. 10-6, the only differences are:

1. i_c lags e by 90°.
2. i_s leads e by the phase angle, ϕ, which is negative.
3. The power factor is now leading.

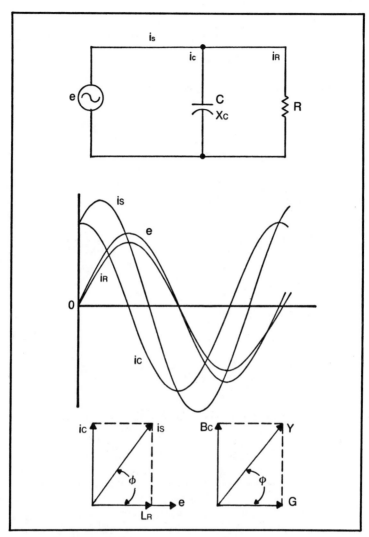

Fig. 10-6. C and R in parallel.

Example 10-6

In Fig. 10-6, C = 35pFs and R = 3.3 k Ω. If the source voltage 50 mV, 1.8 MHz, calculate the values of the total impedance, I_R, I_C, I_S, true power, power factor and phase angle.

Solution

Capacitive Reactance, $X_C = \dfrac{1}{2 \pi fC}$

$$= \frac{1}{2 \times \pi \times 1.8 \times 10^6 \times 35 \times 10^{-12}}$$

$$= 2.53 \text{ k}\Omega$$

Total impedance, $Z = \dfrac{R \times X_C}{\sqrt{R^2 + X_c^2}} = \dfrac{3.3 \times 2.53}{\sqrt{3.3^2 + 2.53^2}}$

$$= 2.0 \text{ k}\Omega$$

$$I_R = \frac{50 \text{ mV}}{3.3 \text{ k}\Omega} = 15.2 \ \mu A.$$

$$I_C = \frac{50 \text{ mV}}{2.53 \text{ k}\Omega} = 19.8 \ \mu A.$$

Supply current, $I_S = \sqrt{I_R^2 + I_C^2} = \sqrt{15.2^2 + 19.8^2}$
$$= 25.0 \ \mu A.$$

Check:

Total impedance $= \dfrac{50 \text{ mV}}{25.0 \ \mu A} = 2.0 \text{ k}\Omega$

True Power $= \dfrac{E^2}{R} = \dfrac{(50 \text{ mV})^2}{3.3 \text{ k}\Omega} = 0.758 \ \mu W.$

Reactive power $= \dfrac{E^2}{X_C} = \dfrac{(50 \text{ mV})^2}{2.53 \text{ k}\Omega} \quad 0.988 \ \mu VAr.$

Apparent power $= E \times I_S = 50 \text{ mV} \times 25 \ \mu A = 1.25 \ \mu VA.$
Power check:
Apparent power $= \sqrt{(\text{True Power})^2 + (\text{Reactive Power})^2}$
$$= \sqrt{0.758^2 + 0.988^2}$$
$$= 1.25 \ \mu VA.$$

Power factor $= \dfrac{\text{True Power}}{\text{Apparent Power}} = \dfrac{0.758}{1.25} = 0.61, \ leading$

Phase angle, $\phi =$ Inv. Cos. 0.61 $= -53°.$

SINEWAVE INPUT TO L AND C IN PARALLEL

Referring to Fig. 10-7, i_C leads e by 90° while i_L lags e by 90°, i_L and i_C are therefore 180° out of phase. Consequently, when i_L and i_C are added together to produce their phase or sum i_S, their magnitudes I_L and I_C will tend to cancel.

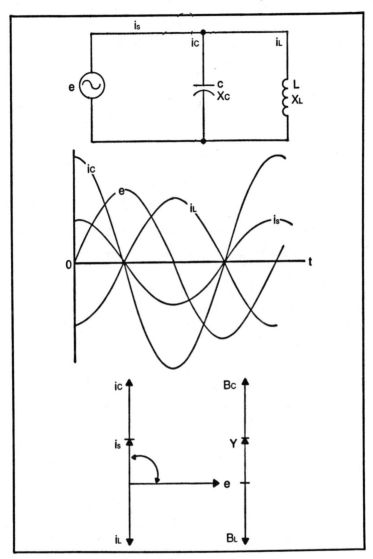

Fig. 10-7. L and C in parallel; X_L is greater than X_C.

If X_L is greater than X_c, then I_L ($= \dfrac{E}{X_L}$) is less than I_C ($= \dfrac{E}{X_C}$) and $I_S = I_C - I_{bl}$. Supply current i_S leads e by 90° ($\phi = -90°$) and the circuit behaves capacitively. This is an opposite result from the equivalent series case, where when X_L was greater than X_C, the circuit behaved inductively.

195

If X_L is less than X_C, then I_L is greater than I_C and $I_S = I_L - I_C$. Supply current i_S lags e by $90°(\phi = +90°)$ and the circuit is inductive.

Whether X_L is greater or less than X_C, the power factor in either case is zero (assuming that there are no resistance losses associated with either L or C).

The total impedance of the circuit, Z is $\dfrac{X_L \times X_C}{X}$, where X is the difference in value between X_L and X_C. Note that the magnitude of Z is always greater than that of either X_L or X_C.

The total admittance of the circuit, $Y = B$, where B is the difference in value between $B_L = \dfrac{1}{X_L}$ and $B_C = \dfrac{1}{X_C}$

If $X_L = X_C$, $I_L = I_C$, $I_S = 0$, Z is theoretically infinite and Y is therefore zero. The parallel combination of L and C then behaves as an open circuit. This situation will also occur at the frequency:

$$f = \frac{1}{2\pi\sqrt{LC}} = \frac{0.159}{\sqrt{LC}} \text{ Hz.}$$

Example 10-7

In Fig. 10-7, $L = 75\mu H$ and $C = 315$ pFs. If the source voltage is 20 mV, 775 kHz, calculate the value of the total reactance, I_L, I_C, I_S and the reactive power.

Solution

Inductive reactance, $X_L = 2\pi fL$
$$= 2 \times \pi \times 775 \times 10^3 \times 75 \times 10^{-6}$$
$$= 365\Omega$$

Capacitive reactance, $X_C = \dfrac{1}{2\pi fC}$

$$= \frac{1}{2 \times \pi \times 775 \times 10 \times 315 \times 10^{-12}}$$
$$= 652\,\Omega.$$

Methods of finding the total reactance:

Method 1

Total reactance, $X = \dfrac{X_L \times X_C}{X_C - X_L} = \dfrac{365 \times 652}{652 - 365}$

196

$$= \frac{365 \times 652}{287}$$

$$= 830 \ \Omega, \text{ rounded off.}$$

Method 2

$$I_L = \frac{E}{X_L} = \frac{20 \text{ mV}}{365 \ \Omega} = 0.0548 \ mA.$$

$$I_C = \frac{E}{X_C} = \frac{20 \text{ mV}}{652 \ \Omega} = 0.0307 \text{ mA.}$$

$$I_S = I_L - I_C = 0.0548 - 0.0307 = 0.0241 \text{ mA.}$$

Since I_L is greater than I_Cn the circuit is inductive.

Total reactance, $X = \dfrac{E}{I_S} = \dfrac{20 \text{ mV}}{0.0241 \text{ mA}} = 830 \ \Omega$ rounded off.

Method 3

Inductive susceptance, $B_L = \dfrac{1}{X_L} = \dfrac{1}{365}$

$$= 2.74 \times 10^{-3} \text{ siemens.}$$

Capacitive susceptance, $B_C = \dfrac{1}{X_C} = \dfrac{1}{652}$

$$= 1.53 \times 10^{-3} \text{ siemens.}$$

Total susceptance, $B = B_L - B_C = 1.21 \times 10^{-3}$ siemens.

Total reactance, $X = \dfrac{1}{B} = \dfrac{1}{1.21 \times 10^{-3}} = 830 \ \Omega$, rounded off.

Total reactive power $= E \times I_S = 10 \text{ mV} \times 0.0241 \text{ mA}$

$$= 0.48 \ \mu VAr, \text{ rounded off}$$

The power factor is zero, lagging, and the phase angle ϕ, is + 90°.

SINEWAVE INPUT TO L, C AND R IN PARALLEL

Referring to Fig. 10-8, current ι_R is in phase with e, ι_C leads e by 90°, and ι_L lags e by 90°; ι_L and ι_C are therefore 180° out of phase so that when they are added together to produce their phasor sum ι_X, their magnitudes, I_L and I_C, will tend to cancel. The phasor sum of ι_R and ι_X is the supply current, ι_S. If X_L is greater than X_C, I_L is less than I_C, and ι_X will be in phase with ι_C. Supply current ι_S leads e by phase angle, ϕ, and the circuit is capacitive. This is the opposite result from the series LCR circuit which was inductive when X_L was greater than X_C. If X_L is less than X_C, ι_X is in phase with ι_S. Supply current ι_S lags e, and the circuit is inductive. The special case of $X_L = X_C$ will be covered in the chapter 11 discussion on resonance.

The total reactance of L and C in parallel is $X_T = \dfrac{X_L \times X_C}{X}$, where X is the difference in value between X_L and X_C. The total impedance is:

$$Z = \frac{RX_T}{\sqrt{R^2 + X_T^2}}$$

which is always less than the value of R; however, Z may be greater than either X_L or X_C (or both X_L and X_C). The supply current is:

$$I_S = \frac{E}{Z} = \sqrt{I_R^2 + I_X^2}$$

where I_X is the difference in value between I_L and I_C

The true power is $\dfrac{E^2}{R} = I_R^2 R = EI_R$ watts.

The apparent power is $\dfrac{E^2}{Z} = I_S^2 Z = EI_S$ volt-amperes.

The power factor is $\dfrac{Z}{R}$ (not $\dfrac{R}{Z}$) so that $\cos\phi = \dfrac{Z}{R}$, $\sin\phi = \dfrac{Z}{X_T}$, $\tan\phi = \dfrac{R}{X_T}$.

The total susceptance, B, of L and C in parallel is the difference in value between $B_L = \dfrac{1}{X_L}$ and $B_C = \dfrac{1}{X_C}$.

Total admittance is $Y = \dfrac{1}{Z} = \sqrt{(G^2 + B^2)}$ siemens, where the conductance is $G = \dfrac{1}{R}$. Then $\cos\phi = \dfrac{G}{Y}$, $\sin\phi = \dfrac{B}{Y}$, $\tan\phi = \dfrac{B}{G}$.

Example 10-8

In Fig. 10-8, R = 2.2 k Ω, L = 3.5 μH, C = 8.5 pFs. If the source voltage is 200 mV, 35 MHZ, calculate the values of the total impedance, I_R, I_L, I_C, I_S, true power, apparent power, power factor and phase angle.

Solution

Inductive reactance, $X_L = 2 \pi fL$
$$= 2 \times \pi \times 35 \times 10^6 \times 3.5 \times 10^{-6}$$
$$= 770 \ \Omega.$$

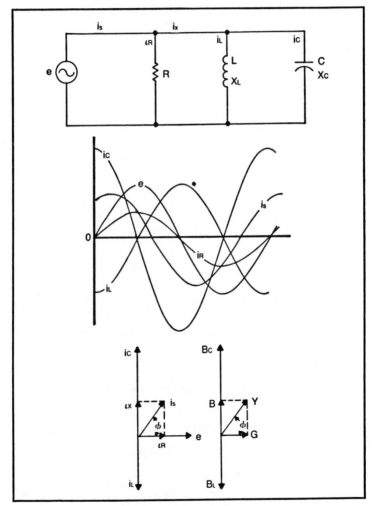

Fig. 10-8. L, C and R in parallel; X_L is greater than X_C.

Capacitive reactance, $X_C = \dfrac{1}{2\pi fC}$

$$= \dfrac{1}{2 \times \pi \times 35 \times 10^6 \times 8.5 \times 10^{-12}}$$

$$= 535 \ \Omega.$$

Total reactance, $X_T = \dfrac{X_L \times X_C}{X_L - X_C} = \dfrac{770 \times 535}{770 - 535}$

$$= \dfrac{770 \times 535 \ \Omega}{235} = 1.75 \ k\Omega.$$

Total impedance $= \dfrac{2.2 \times 1.75}{\sqrt{2.2^2 + 1.75^2}} = 1.37 \ k \ \Omega.$

$I_R = \dfrac{200 \ mV}{2.2 \ k\Omega} = 90.91 \ \mu A.$

$I_L = \dfrac{200 \ mV}{770 \ \Omega} = 259.7 \ \mu A.$

$I_C = \dfrac{200 \ mV}{535 \ \Omega} = 373.8 \ \mu A.$

Supply current, $I_S = \sqrt{90.91^2 + (373.8 - 259.7)^2}$

$$= \sqrt{90.91^2 + 114.1^2}$$

$$= 145.9 \ \mu A.$$

Since I_c is greater than I_L, the line current is capacitive.
Check:

Total impedance, $Z = \dfrac{E}{I_S} = \dfrac{200 \ mV}{145.9 \ \mu A} = 1.37 \ k \ \Omega$

True Power $= \dfrac{E^2}{R} = \dfrac{(200 \ mV)^2}{2.2 \ k \ \Omega} = 18.2 \ \mu W.$

Apparent Power $= E \times I = 200 \ mV \times 149.9 \ \mu A$

$$= 30.0 \ \mu A.$$

Power factor $= \dfrac{\text{True Power}}{\text{Apparent Power}} = \dfrac{18.2}{30.0}$

$$= 0.61, \textit{rounded off.}$$

Check:

Power factor $= \dfrac{Z}{R} = \dfrac{1.37}{2.2} = 0.62.$ *leading*

Phase angle, $\phi =$ Inv. Cos. $0.62 = -52°.$

MAXIMUM POWER TRANSFER-THE AC CASE

With a DC voltage source, there is maximum transfer power to the load when the load resistance is matched (made equal) to the internal resistance of the source. An AC source may possess an internal impedance consisting of resistance and either inductive or capacitive reactance, while the load may be purely resistive or be a combination of resistance and reactance. These two possibilities for the load will be treated separately.

Resistive Load

The magnitude of the source total internal impedance is $\sqrt{(R_S^2 + X_S^2)}$ where X_S may represent either inductive or capacitive reactance (Fig. 10-9). The magnitude of the circuit's total impedance is:

$$Z_T = \sqrt{(R_S + R_\ell)^2 + X_S^2}.$$

Load current, $I_\ell = \dfrac{E}{Z_T} = \dfrac{E}{\sqrt{(R_S + R_\ell)^2 + X_S^2}}$

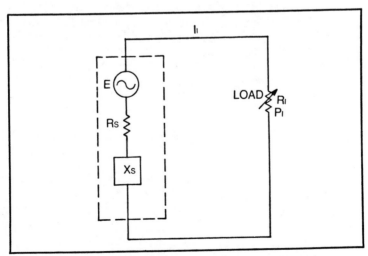

Fig. 10-9. Maximum power transfer with resistive load.

Load power, $P_{\iota} = I^2_{\iota} \times R_{\iota} = \dfrac{E^2_{RL}}{(R_s + R_{\iota})^2 + X_s^2}$

If R_{ι} is varied, it may be shown that the load power, P_{ι}, is a maximum when $R_{\iota} = \sqrt{(R_s^2 \times + X_s^2)}$ or when the load resistance is matched to the magnitude of the source's internal impedance. The value of the maximum power is:

$$P_{\iota(max)} = \dfrac{E^2}{2(R_{\iota} + R_s)} \text{ watts.}$$

Impedance Load

The magnitude of the total impedance of the circuit (Fig. 10-10) is:

$$Z_T = \sqrt{(R_{\iota} + R_s)^2 + (X_{\iota} + X_s)^2}.$$

where X_{ι} and X_s may be either inductive or capacitive. The load current is:

$$I_{\iota} = \dfrac{E}{Z_T} = \dfrac{E}{\sqrt{(R_{\iota} + R_s)^2 + (X_{\iota} + X_s)^2}} \ .$$

The power in the load is:

$$P_{\iota} = I^2 R_{\iota} = \dfrac{E^2 R_{\iota}}{(R_{\iota} + R_s)^2 + (X_{\iota} + X_s)^2} \ .$$

If X_{ι} is varied, P_{ι} will have a maximum value when $X_{\iota} + X_s = 0$, or $X_{\iota} = - X_s$. X_{ι} will then have the same magnitude as X_s but will be opposite in nature so that if, for example X_s is inductive, X_{ι} will be made capacitive, and vice-versa.

After the reactance has been cancelled in the circuit:

$$P_{\iota} = \dfrac{E^2 R_{\iota}}{(R_{\iota} + R_s)^2}.$$

P_{ι} will have a maximum value when R_{ι} is matched to R_s. The value of the maximum power is:

$$P_{\iota(max)} = \dfrac{E^2}{4R_s^2}$$

and the corresponding load current:

$$I_{\iota} = \dfrac{E}{2R_s} \ .$$

With $R_{\iota} = R_S$ and $X_{\iota} = -X_S$, the load impedance and the source internal impedance are said to be conjugates.

Example 10-9

In Fig. 10-9, $R_S = 3$ kΩ, $X_S = 4$ kΩ and E = 12 V. Determine the load power when R_L is (a) 3 kΩ and (b) 7 kΩ. What is the value of R_L for which the load power is a maximum? Calculate the amount of the maximum load power.

Solution

(a)   Total impedance, $Z_T = \sqrt{(3 + 3)^2 + 4^2}$
$$= \sqrt{52} = 7.21 \text{ kΩ}.$$

Load current, $I_{\iota} = \dfrac{E}{Z_T} = \dfrac{12 \text{ V}}{7.21 \text{ kΩ}} = 1.66 \text{ mA}.$

Load power, $P_{\iota} = I_{\iota}^2 \times R_{\iota} = (1.66 \text{ mA})^2 \times 3 \text{ kΩ}$
$$= 8.3 \ mW.$$

(b)   Total impedance, $Z_T = \sqrt{(3 + 7)^2 + 4^2} = \sqrt{116}$
$$= 10.8 \text{ kΩ}.$$

Load current, $I_{\iota} = \dfrac{12 \text{ V}}{10.8 \text{ kΩ}} = 1.11 \text{ mA}.$

Load power, $P_{\iota} = I_{\iota}^2 \times R_{\iota} = (1.11 \text{ mA})^2 \times 7 \text{ kΩ}$
$$= 8.68 \ mW.$$

Load power is a maximum when $R_{\iota} = \sqrt{3^2 + 4^2}$
$$= 5 \text{ kΩ}.$$

Load impedance, $Z_T = \sqrt{(3 + 5)^2 + 4^2} = \sqrt{80}$
$$= 8.94 \text{ kΩ}.$$

Load current, $I_{\iota} = \dfrac{12 \text{ V}}{8.94 \text{ kΩ}} = 1.34 \text{ mA}.$

Maximum load power, $P_{\iota max} = (1.34 \text{ mA})^2 \times 5 \text{ kΩ}$
$$= 8.98 \ mW.$$

Example 10-10

In Fig. 10-10, E = 18 V and $R_S = 4$ kΩ. The source reactance, X_S, is capacitive and equal to 5 kΩ. Determine the values of R_{ι} and X_{ι} which will allow maximum power transfer to the load and find the amount of the maximum load power.

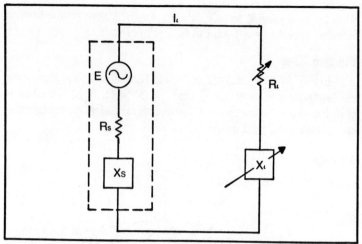

Fig. 10-10. Maximum power transfer with impedance load.

Solution

For maximum power transfer to the load, the load impedance must be the conjugate of the source impedance.

Therefore $R_t = 4 \, k\Omega$ and X_t is an inductive reactance of $5 \, k\Omega$.

$$\text{Maximum load power} = \frac{E^2}{4R_t} = \frac{(18 \text{ V})^2}{4 \times 4 \text{ k}\Omega} = 20.25 \; mW.$$

CHAPTER SUMMARY

☐ *L and R in series.*

$$Z = \sqrt{R^2 + X_L^2} \, \Omega.$$

$$E = IZ, \; V_R = IR, \; V_L = IX_L.$$

$$E = \sqrt{V_R^2 + V_L^2}$$

$$\text{True Power} = I^2R = IV_R = \frac{V_R^2}{R} = E \times I \times \text{p.f watts.}$$

$$\text{Reactive Power} = I^2X_L = IV_L = \frac{V_L^2}{X_L} \; \text{VArs.}$$

$$\text{Apparent Power} = I^2Z = IE = \frac{E^2}{Z} \; \text{VA}$$

$$\text{Apparent Power} = \sqrt{(\text{True Power})^2 + (\text{Reactive Power})^2}$$

$$\text{Power factor, (p.f)} = \frac{R}{Z} = \frac{V^R}{E} = \text{Cos } \phi, \text{ lagging.}$$

ϕ = Inv. Cos. (power factor) and is positive since e leads i.

$$\text{Sin } \phi = \frac{X_L}{Z} = \frac{V_L}{E}, \text{ Tan } \phi = \frac{X_L}{R} = \frac{V_L}{V_R}.$$

☐ *C and R in series.*

$$Z = \sqrt{R^2 + X_C^2} \ \Omega$$
$$E = IZ, \ V_R = IR, \ V_C = IX_C.$$
$$E = \sqrt{V_R^2 + V_C^2}$$

True power = $I^2R = IV_R = \dfrac{V_R^2}{R} = E \times I \times$ p.f. watts.

Reactive power = $I^2X_C = IV_C = \dfrac{V_C^2}{X_C}$ VArs.

Apparent power = $I^2Z = IE = \dfrac{E^2}{Z}$ VA.

Apparent power = $\sqrt{(\text{True Power})^2 + (\text{Reactive Power})^2}$

Power factor = $\dfrac{R}{Z} = \dfrac{V_R}{E} = $ Cos ϕ, leading.

ϕ = Inv Cos (power factor). The sign of ϕ is negative since e leads i.

$$\text{Sin } \phi = \frac{X_C}{Z} = \frac{V_C}{E}, \text{ Tan } \phi = \frac{X_L}{R} = \frac{V_L}{V_R}.$$

☐ *L and C in series.*

$$Z = X_L \sim X_C$$
$$E = IZ, \ V_L = IX_L, \ V_C = IX_C.$$
$$E = V_L \sim V_C$$

True Power = 0 watts.

Reactive Power = $I^2 (X_L \sim X_C) = EI$ VArs.

Power factor = 0 and is leading if X_C is greater than X_L or lagging if X_L is greater than X_C.

ϕ equals + 90° if X_L is greater than X_C but is $-$ 90° if X_C is greater

than X_L. $X_L = X_C$ when the frequency, $f = \dfrac{1}{2 \pi \sqrt{LC}}$ Hz.

□ *L, C and R in series.*

$$Z = \sqrt{R^2 + (X_L \sim X_C)^2}$$

$E = IZ, V_R = IR, V_L = IX_L, V_C = IX_C.$

$$E = \sqrt{V_R^2 + (V_L \sim V_C)^2}$$

True Power $= I^2 R = IV_R = \dfrac{V_R^2}{R} = E \times I \times$ p.f. watts.

Reactive Power $= I^2 (X_L \sim X_C) = I (V_L \sim V_C) = \dfrac{(V_L \sim V_C)^2}{X_{gl} \sim X_C}$ VArs.

Apparent Power $= I^2 Z = IE = \dfrac{E^2}{Z}$ VA

Apparent Power $= \sqrt{(\text{True Power})^2 + (\text{Reactive Power})^2}$

Power factor $= \dfrac{R}{Z} = \dfrac{V_R}{E} = \operatorname{Cos} \phi.$ The power factor is leading if X_C is greater than X_L, but lagging if X_L is greater than X_C.

$\phi =$ Inv. Cos (power factor). The sign of ϕ is negative if X_C is greater than X_L but positive if X_L is greater than X_C.

$$\operatorname{Sin} \phi = \dfrac{X_L \sim X_C}{Z} \ , \ \operatorname{Tan} \phi = \dfrac{X_L \sim X_C}{R}$$

□ *L and R in parallel.*

$$Z = \dfrac{R \times X_L}{\sqrt{R^2 + X_L^2}}$$

$$I_S = \dfrac{E}{Z} \ , \ I_R = \dfrac{E}{R} \ , \ I_L = \dfrac{E}{X_L}$$

$$I_S = \sqrt{I_R^2 + I_L^2}$$

True Power $= \dfrac{E^2}{R} = I_R^2 \times R = E \times I_R = E \times I_S \times$ p.f. watts

Reactive Power $= \dfrac{E^2}{X_L} = I_L^2 \times X_L = E \times I_L$ VArs.

Apparent Power $= \dfrac{E^2}{Z} = I_S^2 \times Z = E \times I_S$ VA.

Apparent Power $= \sqrt{(\text{True Power})^2 + (\text{Reactive Power})^2}$

Power factor $= \dfrac{Z}{R} = \dfrac{I_R}{I_S} = \text{Cos } \phi$, lagging.

ϕ = Inv. Cos, (power factor). ϕ is positive since e leads i_s.

$\text{Sin } \phi = \dfrac{Z}{X_L}$, $\text{Tan } \phi = \dfrac{R}{X_L}$

$B_L = \dfrac{1}{X_L}$, $G = \dfrac{1}{R}$, $Y = \dfrac{1}{Z} = \sqrt{G^2 + B_L{}^2}$ siemens.

□ *C and R in parallel.*

$Z = \dfrac{R \times X_C}{\sqrt{R^2 + X_C{}^2}}$

$I_S = \dfrac{E}{Z}$, $I_R = \dfrac{E}{R}$, $I_C = \dfrac{E}{X_C}$

$I_S = \sqrt{I_R{}^2 + I_C{}^2}$

True Power $= \dfrac{E^2}{R} = I_R{}^2 \times R = E \times I_R = E \times I_S \times \text{p.f.}$ watts

Reactive Power $= \dfrac{E^2}{X_C} = I_C{}^2 \times X_C = E \times I_C$ VArs.

Apparent Power $= \dfrac{E^2}{Z} = I_S{}^2 \times Z = E \times I_S$ VA.

Apparent Power $= \sqrt{(\text{True Power})^2 + (\text{Reactive Power})^2}$

Power Factor $= \dfrac{Z}{R} = \dfrac{I_R}{I_S} = \text{Cos } \phi$, leading.

ϕ = Inv. Cos (power factor). ϕ is negative since e lags i_s

$\text{Sin } \phi = \dfrac{Z}{X_C}$, $\text{Tan } \phi = \dfrac{R}{X_C}$

$B_C = \dfrac{1}{X_C}$, $G = \dfrac{1}{R}$, $Y = \dfrac{1}{Z} = \sqrt{G^2 + B_C{}^2}$ siemens.

□ *L and C in parallel.*

$Z = \dfrac{X_L \times X_C}{X_L \sim X_C}$, $Y = \dfrac{1}{Z} = B_L \sim B_C$

$$I_S = \frac{E}{Z}, \quad I_L = \frac{E}{X_L}, \quad I_C = \frac{E}{X_C}$$

$$I_S = I_L \sim I_C$$

True Power = 0 watts.

Reactive Power = $I^2Z = EI_S$ VArs.

Power factor = 0 and is lagging if X_C is greater than X_L or leading if X_L is greater than X_C.

ϕ equals $+ 90°$ if X_C is greater than X_L but is $- 90°$ if X_L is greater than X_C.

☐ *L, C and R in parallel.*

$$Z = \frac{R \times X}{\sqrt{R^2 + X^2}} \text{ where } X = \frac{X_L \times X_C}{X_L \sim X_C}$$

$$Y = \frac{1}{Z} = \sqrt{G^2 + (B_L \sim B_C)^2}$$

$$I_S = \frac{E}{Z} = EY, \quad I_R = \frac{E}{R} = EG, \quad I_L = \frac{E}{X_L} = EB_L$$

$$I_C = \frac{E}{X_C} = EB_C.$$

$$I_S = \sqrt{I_R^2 + (I_L \sim I_C)^2}$$

True Power = $\frac{E^2}{R} = E^2G = I_R^2 \times R = E \times I_R = E \times I_S \times$ p.f. watts

Total Reactive Power = $\frac{E^2}{X}$ VA$_{rs}$.

Apparent Power = $\frac{E^2}{Z} = I_S^2Z = EI_S$ VA

Apparent Power = $\sqrt{(\text{True Power})^2 + (\text{Total Reactive Power})^2}$

Power Factor = $\frac{Z}{R} = \frac{I_R}{I_S} = \text{Cos } \phi$. The power factor is leading if X_L is greater than X_C but lagging if X_C is greater than X_L.

ϕ = Inv. Cos. (power factor). The sign of ϕ is positive if X_C is greater than X_L but negative if X_L is greater than X_C.

$$\text{Sin } \phi = \frac{Z}{X}, \text{ Tan } \phi = \frac{R}{X}$$

☐ *Maximum Power Transfer-AC Case.*
   Resistive load.

P_ι has its maximum value when $R_\iota = \sqrt{R_S^2 + X_S^2}$

$$P_{\iota(max)} = \frac{E^2}{2\,(R_\iota + R_S)} \text{ watts}$$

☐ *Impedance load.*

P_ι has its maximum value when $R_\iota = R_S$ and $X_L = -X_S$. This means that the load and source impedances are conjugates.

$$P_{\iota(max)} = \frac{E^2}{4\,R_S} = \frac{E^2}{4\,R_\iota}$$

which occurs when the load current, $I_\iota = \frac{E}{2R_\iota} = \frac{E}{2R_S}$.

Resonance

11

There is more than one way in which the condition of resonance may be defined. This book uses the following definition.

Any two-terminal network (this means that there is only one supply) *containing resistance and reactance is said to be in resonance when the supply voltage and the current drawn from the supply are in phase* (the phase angle of the circuit is zero).

It follows from this definition that a resonant circuit has a power factor of unity.

RESONANCE IN A SERIES LCR CIRCUIT

From the definition of resonance, the phase angle ϕ, is zero. Referring to Fig. 11-1, this requires that $X_L = X_C$ and therefore $V_L = V_C$; the phasor sum of v_L and v_C is zero, so that $e = v_R$ and the circuit is purely resistive. The true power is $EI = I^2R = \dfrac{E^2}{R}$ watts.

In terms of magnitudes, $E = V_R$ and $Z = R$. This means that the total impedance of the circuit has a minimum value at resonance and, therefore, the corresponding current, $I = \dfrac{E}{R}$, is at its maximum. For these reasons, the series LCR circuit at resonance is sometimes referred to as an acceptor circuit. With given values of L and C, resonance occurs at a particular frequency:

$$f_r = \frac{1}{2\pi\sqrt{LC}} = \frac{0.159}{\sqrt{LC}}$$

where f_r is measured in Hz, L in henrys and C in farads. Then:

$$L = \frac{1}{4\pi^2 f_r^2 C} = \frac{0.0253}{f_r^2 C} \text{ H, and } C = \frac{1}{4\pi^2 f_r^2 L} = \frac{0.0253}{f_r^2 L} \text{ F.}$$

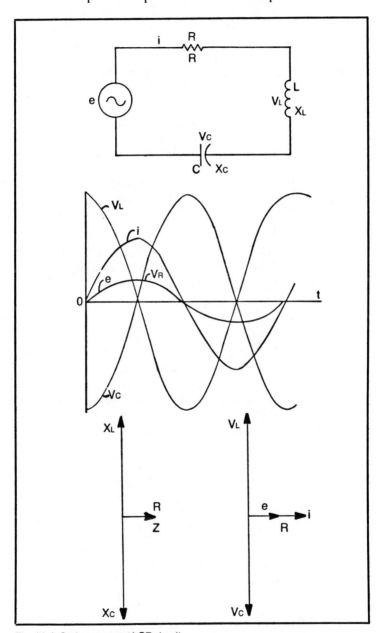

Fig. 11-1. Series resonant LCR circuit.

211

Note that the value of the resonant frequency, f_r, is related to the product of L and C, but is independent of R. Tuning a series LCR circuit means adjusting the value of L or C until the resonant frequency is equal to the desired signal frequency. The behavior of the series LCR circuit as the frequency is varied, is illustrated by means of response curves. These are graphs of particular circuit variables which are plotted against frequency. The most common response curves (Figs. 11-2 A and 11-2 B) are those for the total impedance, Z, and the circuit current, I.

At frequencies below the value of f_r, X_C is greater than X_L, i leads e, and the circuit behaves capacitively. At frequencies above f_r, X_L is greater than X_C, and i lags e; the circuit then behaves inductively.

The quantity "Q" may be regarded as a merit factor related to the inductor in a tuned circuit. It can be defined as the ratio of the inductor's reactance to its resistance so that $Q = \dfrac{X_L}{R}$ and is therefore a number without any units. The Q factor associated with a series tuned circuit has certain interpretations of which the most important follow.

1. Voltage Magnification Factor

At resonance, the circuit current has its maximum value, and equal (but 180° out-of-phase) voltages are developed across the inductor and the capacitor. These voltages may each be many times greater than the applied voltage (refer to example 11-1). The number of times greater is called the magnification factor, which is equal to Q.

Therefore at resonance:

$$Q = \frac{V_L}{E} = \frac{V_C}{E}$$

$$= \frac{IX_L}{IR} = \frac{X_L}{R} = \frac{2\pi f_r L}{R}$$

Since:

$$f_r = \frac{1}{2\pi\sqrt{LC}}$$

$$Q = \frac{2\pi L}{2\pi\sqrt{LC} \times R} = \frac{1}{R} \times \sqrt{\frac{L}{C}}$$

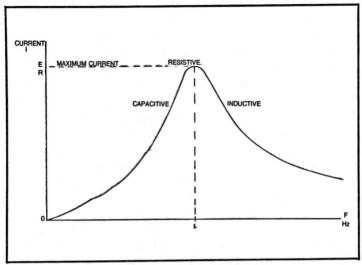

Fig. 11-2 A. Current response curve of the series LCR circuit.

Note that Q is inversely related to R but is directly related to the square root of $\dfrac{L}{C}$. For audio frequency circuits, Q is of the order of 10 but with radio frequency circuits, the value of Q may exceed 100.

2. Reciprocal of the Inductor's Power Factor

Inductor's power factor $= \dfrac{R}{Z}$

$$= \dfrac{R}{\sqrt{R^2 + X_L^{\,2}}}$$

$$= \dfrac{R}{X_L \times \sqrt{1 + \dfrac{R^2}{X_L^{\,2}}}}$$

$$= \dfrac{1}{Q + \sqrt{1 + \dfrac{1}{Q^2}}}$$

which, to within 1%, is equal to $\dfrac{1}{Q}$, provided Q is greater than 10. Therefore, the values of Q and the power factor for radio frequency inductors, are reciprocals.

213

3. Selectivity

The selectivity of a series tuned circuit is defined as its ability to distinguish between the signal frequency to which it is resonant and other signals on nearby frequencies. It therefore follows that the greater the selectivity, the greater is the freedom from adjacent channel interference. The degree of selectivity is related to the sharpness of the current response curve (the sharper the curve, the greater the selectivity) and may be measured by the frequency separation between two specific points on the curve (Fig. 11-3). The points usually chosen are those for which the true power in the circuit is half of the maximum true power which occurs when the circuit is resonant. These positions on the response curve are often referred to as the 3 decibel (dB) points (a loss of 3 dB is equivalent to a power ratio of ½; see appendix A) and their frequency separation is called the bandwidth (or bandpass) value of the tuned circuit. At the 3 dB points, the rms circuit current will be

$$\frac{1}{\sqrt{2}}$$

, or 0.707, times the rms value of the current at resonance (do not confuse this result with the relationship between the rms and the peak values of a sine wave alternating current). In addition, at the 3 dB points, the net (total) circuit reactance is equal to the circuit resistance so that the phase angle is 45° and the power factor is 0.707.

It may be shown that the bandwidth is:

$$f_2 - f_1 = \frac{1}{2 \pi L} \text{ Hz}$$

and therefore:

$$\frac{\text{bandwidth}}{\text{resonant frequency}} = \frac{R}{2 \pi f_r L} = \frac{R}{X_L} = \frac{1}{Q}$$

or: Bandwidth $= \dfrac{\text{resonant frequency}}{Q}$ and Q $= \dfrac{\text{resonant frequency}}{\text{bandwidth}}$.

Therefore Q is a direct measure of the degree of selectivity. The higher the value of Q, the sharper is the current response curve, the greater is the degree of selectivity, and the narrow is the bandwidth.

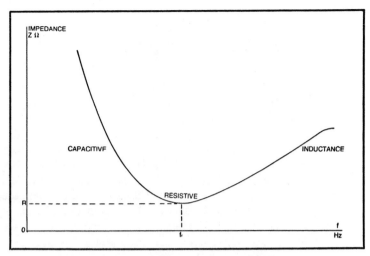

Fig. 11-2 B. Impedance response curve of the series LCR circuit.

4. Energy Relationships

Q is equal to 2π times the ratio of the maximum energy stored during the cycle to the mean energy dissipated over the period.

The maximum energy stored during the cycle in the form of the magnetic field surrounding the inductor, is $\frac{1}{2} L(I_{peak}{}^2)$, where I_{peak} is the peak value of the current at resonance. The maximum energy is therefore $\frac{1}{2}L(\sqrt{2I_{rms}})^2 = L(I_{rms})^2$ joules. The mean energy dissipated over the period T is equal to power × time, or:

Mean energy $= (I_{rms})^2 RT$
$$= (I_{rms})^2 R \times \frac{1}{f_r} \quad \text{joules}$$

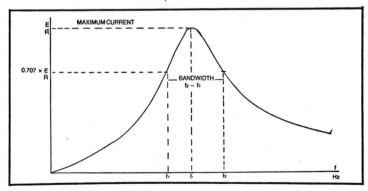

Fig. 11-3. Bandwidth of the series LCR circuit.

215

Then:

$$2\pi \times \frac{\text{maximum energy stored during the cycle}}{\text{mean energy dissipated over the period}}$$

$$= 2\pi \times \frac{L\,(I_{rms})^2}{(I_{rms})^2\,\dfrac{R}{f_r}} = 2\pi \times \frac{f_r\,L}{R} = Q.$$

Note that if two identical series LCR circuits are joined end-to-end with zero mutual coupling between the coils, the total impedance of the series combination at resonance will be doubled ($Z = 2R$), but the Q and the resonant frequency will remain the same. However, if a single additional damping resistor, R_d is connected to the series LCR circuit, the new impedance at resonance is greater, and equal to $R + R_d$ so that the new Q is given by:

$$Q_{new} = \frac{1}{R + R_d} \times \sqrt{\frac{L}{C}} = \frac{R}{R + R_d} \times Q_{old}$$

and is reduced. New bandwidth $= \dfrac{(R_T + R_d)}{R} \times$ old bandwidth, and is increased, but the resonant frequency is unchanged.

Example 11-1

In Fig. 11-1, $R = 8\,\Omega$, $L = 150\,\mu H$, $C = 250\,pFs$ and $E = 1.4\,V$. Calculate the resonant frequency and determine the resonant values of Z, I, V_R, V_L, V_C and the true power. What are the values of Q and the circuit's bandwidth? If an additional $12\,\Omega$ resistor is inserted in series, calculate the new values of the resonant frequency, Q and bandwidth.

Solution

The resonant frequency: $f_r = \dfrac{1}{2\pi\sqrt{LC}}$

$$= \frac{0.159}{\sqrt{150 \times 10^{-6} \times 250 \times 10^{-12}}}\ \text{Hz}$$

$$=\quad 822\ \text{kHz}.$$

At the resonant frequency: $X_L = X_C = 2 \times \pi \times 822 \times 10^3$
$$\times 150 \times 10^{-6}$$
$$= 775\Omega.$$

At resonance:

$$Z = R = 8 \ \Omega.$$

$$I = \frac{E}{R} = \frac{1.4 \ V}{8\Omega} = 175 \ mA.$$

$V_R = E = 1.4 \ V.$
$V_L = IX_L = 175 \, mA \times 775\Omega = 136V.$
$V_C = V_L = 136 \ V.$

True Power $= I^2R = (175 \ mA)^2 \times 8 \ \Omega = 0.245 \ W.$
The power factor is unity and the phase angle between
e and i is zero.

$$Q = \frac{1}{R} \times \sqrt{\frac{L}{C}} = \frac{1}{8} \times \sqrt{\frac{150 \times 10^{-6}}{225 \times 10^{-12}}} = 97.$$

Check: $Q = \dfrac{X_L}{R} = \dfrac{X_C}{R} = \dfrac{775 \ \Omega}{8} = 97.$

and $Q = \dfrac{V_L}{E} = \dfrac{v_C}{E} = \dfrac{136 \ V}{1.4 \ V} = 97.$

Bandwidth $= \dfrac{R}{2 \ \pi \ L} = \dfrac{8}{2 \times \pi \times 150 \times 10^{-6}}$ Hz $= 8.5 \ kHz$

Check: Bandwidth $= \dfrac{f_r}{Q} = \dfrac{822 \ kHz}{97} = 8.5 \ kHz.$

When the additional 12 Ω resistor is inserted in series, the re-
sonant frequency is unchanged. The new Q is $97 \times \dfrac{8 \ \Omega}{8\Omega + 12 \ \Omega}$
$= 39$ and the new bandwidth is $\dfrac{822}{39} = 21.2 \ kHz.$

RESONANCE IN A PARALLEL LCR CIRCUIT

The phase relationship between e, i_R, i_L and i_C have already
been described. At resonance the phase angle, ϕ, between e and i_s,
is zero; This requires that $I_L = I_C$, and $I_S = I_R$ (Fig. 11-4). The total
impedance of the circuit, Z, is resistive and also has a maximum

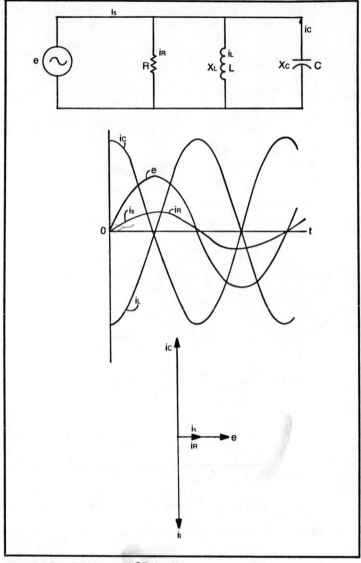

Fig. 11-4. Parallel resonant LCR circuit.

value which is equal to R (note that the circuit impedance can never exceed R). The supply current, I_s, is a minimum and equal to $\dfrac{E}{R}$. The resonant frequency is: $f_r = \dfrac{1}{2 \pi \sqrt{LC}} = \dfrac{0.159}{\sqrt{LC}}$ Hz which is the same expression as for series resonance.

218

Then: $L = \dfrac{1}{4\,\pi^2 f_r^2 C} = \dfrac{0.0253}{f_r^2 C}$ H

and $C = \dfrac{1}{4\,\pi^2 f_r^2 L} = \dfrac{0.0253}{f_r^2 L}$ F.

The response curves for the parallel LCR circuit are shown in Figs. 11-5A and 11-5B.

At frequencies below the resonant frequency f_r, X_C is greater than X_L, I_C is less than I_L and the circuit behaves inductively. At frequencies above the resonance frequency, X_L is greater than X_C, I_L is less than I_C and the circuit behaves capacitively. These results are the reverse to those for the series LCR circuit, which behaved capacitively for frequencies below f_r and inductively for frequencies above f_r.

The α factor may be interpreted as:

1. Current Magnification

At resonance, the currents I_L and I_C are equal in magnitude, and each is Q times the supply current, I_s.

Then: $I_L = \dfrac{E}{X_L}$, $I_C = \dfrac{E}{X_C}$, $I_s = I_R = \dfrac{E}{R}$

$Q = \dfrac{I_L}{I_s} = \dfrac{\dfrac{E}{X_L}}{\dfrac{E}{R}} = \dfrac{R}{X_L} = \dfrac{R}{X_C}$ *(not* $\dfrac{X_L}{R}$ *or* $\dfrac{X_C}{R}$

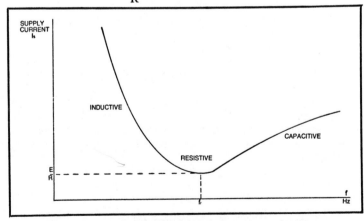

Fig. 11-5 A. Current response curve of the parallel resonant LCR circuit.

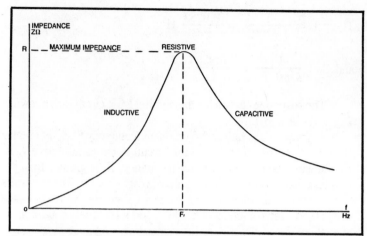

Fig. 11-5B. Impedance response curve of the parallel resonant LCR circuit.

Since: $Q = \dfrac{R}{X_L} = \dfrac{R}{2\pi f_r L}$ and $f_r = \dfrac{1}{2\pi\sqrt{LC}}$

$$Q = \dfrac{R}{2\pi L/2\pi\sqrt{LC}} = R \times \sqrt{\dfrac{C}{L}}$$

2. Impedance Magnification

Since $R = QX_L = QX_C$, the impedance at resonance is Q times the reactance of either the inductor or the capacitor.

3. Selectivity

As in the case in the series LCR circuit, Q determines the sharpness of the response curves and is therefore a direct measure of selectivity.

$$Q = \dfrac{\text{resonant frequency}}{\text{bandwidth}}$$

and the bandwidth may be defined for the impedance response curve as the frequency separation between the total impedance of the current is $0.707 \times$ the maximum impedance, R; from this definition:

$$\text{Bandwidth} = \dfrac{1}{2\pi CR}\text{ Hz}$$

220

where C is measured in farads and R in ohms. The concept of the half-power points cannot be used in the case of the parallel LCR circuit since the true power is always $\dfrac{E^2}{R}$ and is independent of frequency.

4. Energy Relationships

As in the case in the series LCR circuit, Q may be defined as:

$$Q = 2\pi \times \frac{\text{maximum energy stored during the cycle}}{\text{mean energy dissipated during the period}}$$

The maximum energy stored in the capacitor is:

$$\frac{1}{2} C(E_{peak})^2 = \frac{1}{2}C (\sqrt{2E_{rms}})^2 = C(E_{rms})^2.$$

The mean energy dissipated during the period is

$$\frac{(E_{rms})^2}{R} \times T$$

where T is the period which is equal to $\dfrac{1}{f_r}$

Then: $Q = 2\pi \times \dfrac{C(E_{rms})^2}{T\,(E_{rms})^2/R} = R \times 2\pi\, f_r C = \dfrac{R}{X_C}$

Note that if two identical parallel LCR circuits are shunted across each other with zero mutual coupling between the coils, the effective resistance is halved. The impedance at resonance is therefore halved but the Q, the resonant frequency, and the bandwidth remain the same.

If a single damping resistor, R_d, is connected across the parallel LCR circuit, the new impedance at resonance is:

$$\frac{R_d \times R}{R + R_d}$$

and is reduced.
Then:

$$Q_{new} = \frac{R_d \times R}{R + R_d} \times \sqrt{\frac{C}{L}} = \frac{R_d}{R + R_d} \times Q_{old}$$

and the value of Q has therefore been decreased.

$$\text{New Bandwidth} = \frac{R + R_d}{R_d} \times \text{old bandwidth}$$

and is consequently increased, but the resonant frequency remains the same.

Example 11-2

In Fig. 11-4, R = 8.2 kΩ, L = 3.3 μH, C = 6.5 pFs and E = 15V. Calculate the resonant frequency and determine the resonant values of I_s, I_R, I_L, I_C and the true power. What are the values of Q and the bandwidth of the circuit. If an additional 10 kΩ resistor is added in parallel what are the new values of the resonant frequency, Q and bandwidth?

Solution

The resonant frequency, $f_r = \dfrac{1}{2\pi\sqrt{LC}}$

$$= \frac{1}{2 \times \pi \times \sqrt{3.3 \times 10^{-6} \times 6.5 \times 10^{-12}}} \ H$$

$$= 34.4 \ MHz.$$

At resonance, the inductive reactance, $X_L = X_C$

$$= 2 \times \pi \times 34.4 \times 10^6 \times 3.3 \times 10^{-6} = 712.5 \ \Omega.$$

$$I_R = I_S = \frac{E}{R} = \frac{15V}{8.2 \ k\Omega} = 1.83 \ mA.$$

$$I_L = I_C = \frac{15 \ V}{712.5 \ \Omega} = 21.05 \ mA.$$

$$\text{True Power} = \frac{E^2}{R} = \frac{(15 \ V)^2}{8.2 \ k\Omega} = 27.4 \ mW.$$

$$Q = R \times \sqrt{\frac{C}{L}} = 8.2 \times 10^3 \times \sqrt{\frac{6.5 \times 10^{-12}}{3.3 \times 10^{-6}}} = 11.5.$$

Check:

$$Q = \frac{I_c}{I_s} = \frac{R}{X_C} = \frac{8.2 \times 10^3}{712.5} = 11.5.$$

$$\text{Bandwidth} = \frac{1}{2\pi CR} = \frac{1}{2 \times \pi \times 6.5 \times 10^{-12} \times 8.2 \times 10^3} \ Hz$$

$$= 2.99 \ MHz.$$

Check:

$$\text{Bandwidth} = \frac{f}{Q} = \frac{34.4 \text{ MHz}}{11.5} = 2.99 \text{ MHz}.$$

When the additional 10 k Ω resistor is added in parallel, the new total resistance is $\frac{10 \times 8.2}{10 + 8.2} = 4.5 \text{ k}\Omega$. The resonant frequency remains unchanged but the new Q is $11.5 \times \frac{4.5}{8.2} = 6.3$ and the new bandwidth is $2.99 \times \frac{8.2}{4.5} = 5.45 \text{ MHz}.$

PARALLEL RESONANT TANK CIRCUIT

This circuit consists of a practical coil with resistance, in parallel with a capacitor whose losses are assumed to be negligible (Fig. 11-6). Such a circuit is commonly used as the plate or collector load of certain rf amplifiers.

Since the inductor branch contains both inductive reactance and resistance, current ι_L lags e by a phase angle, ϕ. If the frequency is raised, the impedance of the inductor branch will increase, ι will decrease and the angle, ϕ, will approach 90°.

Capacitor current i_c leads e by 90°, and supply current i_s is the phasor sum of i_L and i_c. Figure 11-7 illustrates changes in the circuit behavior as the frequency is varied. At all frequencies below the resonant frequency f_r, i_s lags e by the total phase angle, ϕ, and the circuit behaves inductively. At all frequencies above f_r, i_s leads

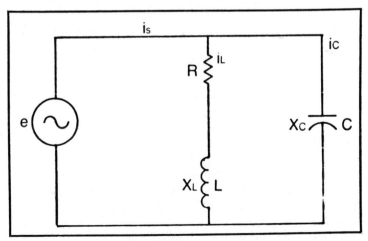

Fig. 11-6. Parallel resonant tank circuit.

223

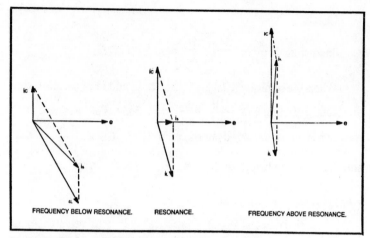

FREQUENCY BELOW RESONANCE.　　RESONANCE.　　FREQUENCY ABOVE RESONANCE.

Fig. 11-7. Phasor diagram of the parallel tank circuit.

e and the circuit is capacitive. These results are similar to those obtained for the parallel LCR circuit discussed in chapter eleven but are opposite to those for the series LCR circuit discussed earlier in that chapter.

At resonance, e and i_s are in phase so that $\theta = 0°$. For resonant tank circuits possessing a high Q, I_L may be considered as equal to I_C and X_L as equal to X_C. However, the exact expression for the resonant frequency, f_r, is :

$$f_r = \frac{1}{2\pi} \times \sqrt{\frac{1}{LC} - \frac{R^2}{L^2}} \quad Hz.$$

For rf tank circuits possessing an appreciable Q (greater than 10), the

$$\frac{R^2}{L^2}$$

term is much less in value than the

$$\frac{1}{LC}$$

term. Therefore, f_r approximately equals

$$\frac{1}{2\pi \times \sqrt{LC}}$$

which is the same as the expression for f_r in chapter eleven. As suggested by the phasor diagram in Fig. 11-7, the supply current I_s is small at resonance and is nearly equal to the minimum value possible. The total impedance at resonance is therefore close to

Humber Libraries
library.humber.ca
(416) 675-5079

Library name: LAKESHORE
User name: Al-Ali, Iman

Title: AC/DC electricity and
electronics made easy
Call number: QC522 .V44
Date charged: 9/26/2018,12:
Date due: 10/10/2018,23:59

Total checkouts for session:1
Total checkouts:1

If items are not returned or
renewed before the due date,
you will be charged fines.

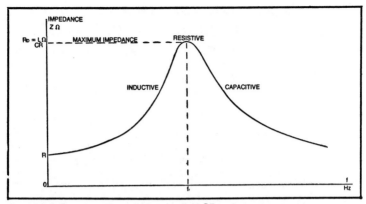

Fig. 11-8 A. Response curve of the parallel LCR tank circuit.

the maximum value and is equal to

$$\frac{L}{CR} \, \Omega.$$

From the definition of resonance, this quantity is resistive and is called the dynamic resistance; the term "dynamic" means that "L

$$" \frac{L}{CR} \, \Omega"$$

only appears under operating conditions. The parallel tank is sometimes referred to as a rejector circuit since it presents virtually maximum impedance at resonance. The principal resonance curves of the tank circuit are shown in Figs. 11-8A and 11-8B.

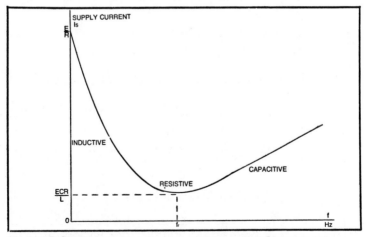

Fig. 11-8 B. Supply current response curve of the parallel LCR tank circuit.

The Q factor may be interpreted as:

1. **Impedance Magnification**—The impedance at resonance is Q times the reactance of the capacitor branch. Therefore:

$$Q = \frac{L/(CR)}{1/(2\pi f_r C)} = \frac{2\pi f_r L}{R} = \frac{X_L}{R}$$

which is the same expression obtained for the series LCR resonant circuit.

Since X_L and X_C are nearly equal at resonance, $Q = \frac{X_C}{R}$, and for the impedance at resonance, $Z = X_C Q \approx Q^2 R$ (the symbol "$\approx$" means" is approximately equal to").

2. **Current Magnification**—Since i_L and i_C are nearly equal in magnitude and approximately 180° out of phase, there appears to be a large circulating, or "flywheel," current existing between L and C. The supply, or "make-up," current, I_s, is relatively small since it merely accounts for the fact that i_L and i_C are not exactly equal in magnitude and 180° out of phase. The ideal case of R = 0 was covered earlier, which showed that when $X_L = X_C$, the total impedance and Q were both infinite, and the supply current I_s was zero.

In the practical case, the impedance at resonance is Q times the reactance of the capacitor branch and is approximately equal to Q times the impedance of the inductor branch. Therefore the circulating current is Q times the supply current.

The true power in the circuit at resonance is $I_s^2 \times$ the dynamic resistance $= I_L^2 R$, or:

$$\frac{\text{circulating current, } I_L}{\text{make-up current, } I_s} = \sqrt{\frac{L/(CR)}{R}} = \frac{1}{R} \times \sqrt{\frac{L}{C}} = Q$$

3. **Selectivity**—Q is again a measure of selectivity and equals $\frac{f_r}{\text{bandwidth}}$. Using the impedance response curve, the bandwidth may be defined as the frequency separation between the points where the total circuit impedance is 0.707 times the impedance at resonance.

4. **Energy Relationships**—Here again

$$Q = 2\pi \times \frac{\text{maximum energy stored during the cycle}}{\text{mean energy disipated over the period}}$$

Note that if two indentical tank circuits are paralleled with zero mutual coupling between the coils, the impedance at resonance is halved, but the Q and the resonant frequency remain unchanged. If a single damping resistor, R_d, is connected across the tank circuit, the new impedance at resonance is:

$$Z_{new} = \frac{R_d \times L/(CR)}{R_d + L/(CR)}$$

$$\text{and } Q_{new} = \frac{R_d}{R_d + L/(CR)} \times Q_{old}$$

so that both quantities are reduced. The bandwidth is therefore greater but the resonant frequency remains the same(assuming the Q is greater than 10).

Example 11-3

In Fig. 11-6, $R = 18\ \Omega$, $L = 25\ \mu H$, $C = 15pFs$ and $E = 100$ mV. Determine the resonant frequency and calculate the resonant values of I_S, I_L, I_C. What are the values of Q and the tank circuit bandwidth?

Solution

The exact formula for the resonant frequency is:

$$f_r = \frac{1}{2\pi} \times \sqrt{\frac{1}{LC} - \frac{R^2}{L^2}} \ .$$

However: $\frac{1}{LC} = \frac{1}{25 \times 10^{-6} \times 15 \times 10^{-12}} = 2.667 \times 10^{15}$

and $\frac{R^2}{L^2} = \frac{324}{(25 \times 10^{-6})^2} = 5.18 \times 10^{11}$.

Therefore: $\frac{R^2}{L^2}$ is negligible when compared with $\frac{1}{LC}$.

The formula then becomes:

$$f_r = \frac{1}{2\pi \times \sqrt{LC}} = \frac{1}{2 \times \pi \times \sqrt{25 \times 10^{-6} \times 15 \times 10^{-12}}}$$

$$= 8.22 \text{ MHz}$$

At the resonant frequency, $X_C = \dfrac{1}{2 \pi f_r C}$

$$= \dfrac{1}{2 \times \pi \times 8.22 \times 10^6 \times 15 \times 10^{-12}}$$

$$= 1291 \ \Omega$$

The impedance of the inductor branch is $\sqrt{R^2 + X_L{}^2} =$ $\sqrt{18^2 + 1291^2} \approx 1291 \ \Omega$.

The circulating current, $I_L = I_C = \dfrac{100 \ mV}{1291 \ \Omega} = 0.07746 \ mA$

$$= \textit{77.46} \ \mu A.$$

Dynamic resistance, $R_D = \dfrac{L}{CR} = \dfrac{25 \times 10^{-6}}{15 \times 10^{-12} \times 18} \ \Omega$

$$= 92.6 \ k\Omega.$$

Supply or make-up current, $I_S = \dfrac{100 \ mV}{92.6 \ k\Omega} = \textit{1.08} \ \mu A.$

$Q = \dfrac{X_L}{R} = \dfrac{1291}{18} = \textit{71.7}$

Check: $Q = \dfrac{I_C}{I_S} = \dfrac{77.46}{1.08} = \textit{71.7}$

Bandwidth $= \dfrac{f_r}{Q} = \dfrac{8.22}{71.7} \ MHz = \textit{115} \ kHz.$

CHAPTER SUMMARY

☐ *Series Resonant LCR Circuit.*
$\phi = 0°$. Power factor is unity.

Resonant Frequency, $f_r = \dfrac{1}{2 \pi \times \sqrt{LC}} \ Hz.$

$$L = \dfrac{0.0253}{f_r{}^2 C} H, \quad C = \dfrac{0.0153}{f_r{}^2 L} \ F.$$

$Z = R$ (minimum value), $I = \dfrac{E}{R}$ (maximum value).

$$V_R = E, \; V_L = V_C = Q \times E,$$

$$Q = \frac{V_L}{E} = \frac{V_C}{E} = \frac{X_L}{R} = \frac{X_C}{R} = \frac{1}{R} \times \sqrt{\frac{L}{C}}$$

$$= \frac{1}{\text{inductor's power factor}}$$

$$\text{Bandwidth} = \frac{R}{2 \pi L} \text{ Hz} = \frac{f_r}{Q}$$

Current at 3 dB points = 0.707 × resonant current.
Maximum power at resonance = I^2R watts.
□ *Parallel Resonant LCR Circuit.*
Circuit phase angle is zero. Power factor is unity.

Resonant frequency, $f_r = \dfrac{1}{2 \pi \sqrt{LC}}$ Hz.

$Z = R$ (maximum value), $I_s = \dfrac{E}{R}$ (minimum value).

$I_L = I_C = Q \times I_s.$

$$Q = \frac{I_L}{I_s} = \frac{I_C}{I_s} = \frac{R}{X_L} = \frac{R}{X_C} = R \times \sqrt{\frac{C}{L}}$$

$$\text{Bandwidth} = \frac{f_r}{Q} = \frac{1}{2 \pi RC} \text{ Hz}$$

Impedance at bandwidth points = 0.707 × R.
□ *Parallel Resonant Tank Circuit.*
Circuit phase angle is zero. Power factor is unity.

Resonant frequency, $f_r = \dfrac{1}{2 \pi} \sqrt{\dfrac{1}{LC} - \dfrac{R^2}{L^2}} \approx \dfrac{1}{2 \pi \sqrt{LC}}$ Hz

provided Q is greater than 10.

$$Z = R_D = \frac{L}{CR} = Q \times X_C = Q \times X_L \approx Q^2R \; \Omega. \; I_s = \frac{ECR}{L}$$

$$Q = \frac{X_L}{R} = \frac{R_D}{X_L} = \frac{R_D}{X_C} = \frac{I_L}{I_s} = \frac{I_C}{I_s} = \frac{1}{R} \times \sqrt{\frac{L}{C}}$$

$$\text{Bandwidth} = \frac{f_r}{Q}$$

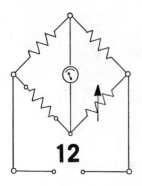

Mutually

Coupled Circuits

12

Previous chapters have referred to those inductors having a magnetic flux linked with their own turns, therefore possessing the property of self-inductance. We are now going to consider the case where the magnetic flux associated with one coil links with the turns of another coil. Such inductors are said to be magnetically, inductively, mutually, or transformer coupled and possess the property of mutual inductance, M.

MUTUAL INDUCTANCE

Alternating current i_1 in the coil 1 of Fig. 12-1 creates an alternating magnetic flux, part of which only links coil 1 (this is referred to as the leakage flux), and the remainder links coil 2. This causes an induced voltage, v_2, whose magnitude in part, depends on i_1 and the value of the mutual inductance between the coils. Mutual inductance, M, like self-inductance, is measured in henrys; the mutual inductance is 1 henry if, when the current i_1 is instantaneously changing at the rate of 1 ampere per second, the induced voltage, v_2 is 1 volt. The factors that determine the value of the mutual inductance include the number of turns in N_1 and N_2, the cross-sectional area of the coils, their separation, the orientation of their axes, and the nature of their cores.

Induced voltage v_2 = M × rate of change of i_1. In terms of AC sinewave values, $V_{2rms} = 2\pi fMI_{1rms}$ where f is the frequency in hertz of the current i_1 (compare $V_{Lrms} = 2\pi FLI_{rms}$ for the property of self-inductance). Provided the two coils are wound in the same sense, v_2 lags i_1 by 90° but if the coils are in the opposite sense, v_2 leads i_1 by 90°. Note that the property of mutual inductance is reversible so that if the same rate of change of the current, i_1, is flowing in coil 2, then the voltage induced in coil 1 is v_1 = M × rate of change of i_1.

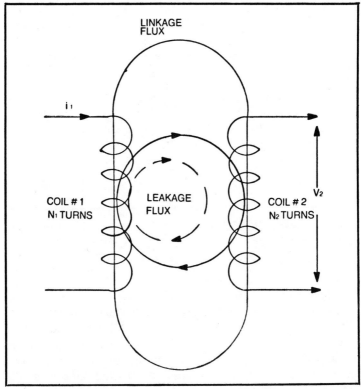

Fig. 12-1. Mutual Coupling.

In the extreme case where the two coils are tightly wound, one on top of the other, with a common soft iron core, the leakage flux is extremely small and can be neglected. Assuming perfect flux linkage between the coils (corresponding to zero leakage flux), the mutual inductance is:

$$M = \frac{\mu_o \mu_r N_1 N_2 A}{\iota} \text{ henrys}$$

where A is the cross-sectional area of each of the coils in squaremeters, μr is the relative permeability of the soft iron, ι is the length of the coils in meters and $\mu_0 = 4\mu_r \times 10^{-7}$ SI units.

Since the self-inductance of the coil 1 is

$$L_1 = \frac{\mu_0 \mu r N_1^2 A}{\iota} \text{ henrys}$$

and, in a similar way:

$$L_2 = \frac{\mu_0 \mu r N_2^2 A}{\iota} \text{ henrys}$$

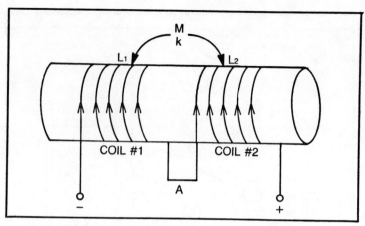

Fig. 12-2 A. Coils connected in series-aiding.

Then:

$$M^2 = L_1 L_2 \text{ and } M = \sqrt{L_1 L_2}$$

THE COUPLING FACTOR

If the linkage flux is not negligible, then only a fraction, k, of the total flux links the two coils. This fraction k, which cannot exceed 1, is called the coefficient of coupling, or coupling factor; its value can be close to 1 if a common soft iron core is used for the two coils, but may be very small (less than 0.01) with an air core and the coils widely separated.

It may be shown that:

$$k = \frac{M}{\sqrt{L_1 L_2}} \text{ or } M = k \times \sqrt{L_1 L_2}$$

Then:

$$L_1 = \frac{M^2}{k^2 L_2} \text{ and } L_2 = \frac{M^2}{k^2 L_1}$$

Also:

$$M = kL_1 \times \frac{N_2}{N_1} = kL_2 \times \frac{N_1}{N_2}$$

If L_1 and L_2 are equal to L:

$$k = \frac{M}{L} \text{ and } M = kL$$

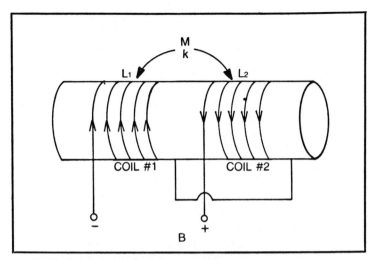

Fig. 12-2 B. Coils connected in series-opposing.

Note that if a steady direct current is flowing through coil 1, the linkage flux will be constant in magnitude and direction so that the voltage induced in the coil 2, will be zero.

Mutually Coupled Coils in Series

If two mutually coupled coils having a common axis are connected in series, their individual fluxes may aid or oppose, depending on the sense in which the coils are wound.

The direction of the arrows, indicative of current flow, clearly demonstrate that in Fig. 12-2 A, the individual fluxes surrounding L_1 and L_2 will be aiding while in Fig. 12-2 B, they will be opposing.

First consider the case in which the fluxes are aiding, as shown in Fig. 12-3.

Let the rate of change of current through the series circuit be I/t amperes per second. The voltage induced in coil 1 will be L_1I/t volts due to self-inductance and MI/t volts due to the flux surrounding coil 2 and linking partially with coil 1; likewise the voltage induced in coil 2 will be L_2I/t from self-inductance and MI/t from

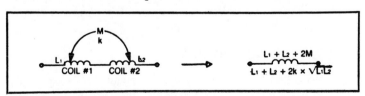

Fig. 12-3. Coils in series-aiding.

mutual inductance. Since the fluxes are aiding, the total voltage across the series circuit will be:

$$(L_1 I/t + MI/t) + (L_2 I/t + MI/t) = (L_1 + L_2 + 2M)I/t$$

Then the total inductance, L_T, will be $L_1 + L_2 + 2M$, as shown in the equivalent circuit of Fig. 12-3.

If the fluxes are opposing, the sign of M is reversed, and the total inductance becomes $L_T = L_1 + L_2 - 2M$.

The equations developed are:

Series-aiding coils:

$$L_T{}^+ = L_1 + L_2 + 2M = L_1 + L_2 + 2k \times \sqrt{L_1 L_2}$$

Series-opposing coils:

$$L_T{}^- = L_1 + L_2 - 2M = L_1 + L_2 - 2k \times \sqrt{L_1 L_2}$$

Therefore:

$$L_1 + L_2 = \frac{L_T{}^+ + L_T{}^-}{2}$$

and

$$M = \frac{L_T{}^+ - L_T{}^-}{4}$$

The positive sign used with L_T means "aiding", the negative sign means "opposing".

If L_1 and L_2 are each equal to L, the equations become:

Series-aiding coils:

$$L_T{}^+ = 2(L + M) = 2L(1 + k)$$

Series-opposing coils:

$$L_T{}^- = 2(L - M) = 2L(1 - k)$$

Then:

$$L = \frac{L_T{}^+ + L_T{}^-}{4} \ , \ M = \frac{L_T{}^+ - L_T{}^-}{4} \quad k = \frac{L_T{}^+ - L_T{}^-}{L_T{}^+ + L_T{}^-}$$

Mutually Coupled Coils in Parallel

Figure 12-4 represents two mutually coupled coils in parallel. If the sense of the inductors is such that the coils are in parallel-

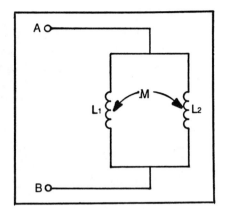

Fig. 12-4. Parallel coils with mutual inductance.

aiding, the total inductance, L_T^+, between points A and B is:

$$L_T^+ = \frac{L_1L_2 - M^2}{L_1 + L_2 + 2M}$$

However, if the coils are connected in parallel-opposing, the sign of M is reversed and the total inductance is:

$$L_T = \frac{L_1L_2 - M^2}{L_1 + L_2 + 2M}$$

(The above expression for L_T^+ and L_1^- are not the same as those derived from:

$$\frac{1}{L_T} = \frac{1}{L_1 \pm M} + \frac{1}{L_2 \pm M}$$

(which is erroneously shown in some textbooks).

Note that if $M = 0$, both equations become

$$L_T = \frac{L_1L_2}{L_1 + L_2}$$

which is familiar product-over-sum formula for self-inductances.

If L_1 and L_2 are each equal to L, the equations become:

$$L_T^+ = \frac{L^2 - M^2}{2(L - M)} = \frac{L + M}{2}$$

and

$$L_T^- = \frac{L - M}{2}.$$

Example 12-1

Two mutually coupled coils are joined in a series-aiding arrangement. If the self-inductances are 0.75 H, 0.6 H and the coupling factor is 0.8, find the total equivalent inductance. If one of the coils is now reversed without changing the coupling factor, what is the new value of the total equivalent inductance?

Solution

Mutual inductance, $M = k \times \sqrt{L_1 L_2} = 0.8 \times \sqrt{0.75 \times 0.6}$
$$= 0.537 \text{ H}.$$

Total equivalent inductance, $L_T^+ = L_1 + L_2 + 2M$
$$= 2.42 \text{ H}.$$

If one of the coils is reversed, the connection is now series-opposing.

Total equivalent inductance, $L_T^- = L_1 + L_2 - 2M$
$$= 0.276 \text{ H}.$$

Example 12-2

Two coils whose self inductances are $75\mu H$ and $125\mu H$, have a mutual inductance of 15 μH. What is the couplign factor? Calculate the equivalent inductance if the coils are connected in (a) series-aiding and (b) series-opposing?

Solution

Coupling factor, $k = \dfrac{M}{\sqrt{L_1 L_2}} = \dfrac{15}{\sqrt{75 \times 125}}$
$$= 0.155.$$

(a) In the series-aiding connection:

$$L_T^+ = L_1 + L_2 + 2M = 75 + 125 + 2 \times 15$$
$$= 230 \,\mu H.$$

(b) In the series-opposing connection:

$$L_T^- = L_1 + L_2 - 2M = 75 + 125 - 2 \times 15$$
$$= 170 \,\mu H.$$

Example 12-3

Two coils whose self-inductances are 65 mH and 85 mH are connecting in parallel-aiding, with a coupling factor of 0.35. What is the total equivalent inductance of the parallel combination? If one of

236

the coils is now reversed without changing the coupling factor, what is the new value of the total equivalent inductance?

Solution

Mutual inductance, $M = k \times \sqrt{L_1 L_2}$
$$= 0.35 \times \sqrt{65 \times 85}$$
$$= 26 \text{ mH.}$$

Parallel-aiding connection, $L_T^+ = \dfrac{L_1 L_2 - M^2}{L_1 + L_2 - 2M}$

$$= \dfrac{65 \times 85 - 26^2}{65 + 85 - 2 \times 26}$$

$$= 49.5 \text{ mH.}$$

Parallel-opposing connection, $L_T^- = \dfrac{L_1 L_2 - M^2}{L_1 + L_2 + 2M}$

$$= \dfrac{65 \times 85 - 26^2}{65 + 85 + 2 \times 26}$$

$$= 24.0 \text{ mH}$$

POWER TRANSFORMERS

A power transformer consists of primary and secondary coils whose numbers of turns are N_p and N_s, Fig. 12-5 A. These are wound on a common soft-iron core which reduces the leakage flux to a low value. In the case of the ideal transformer, the leakage flux is zero so that the mutual inductance, $M = \sqrt{L_p L_s}$, and the coupling factor, k, is unity. When an alternating current, I_p (rms), flows in the primary coil, it creates a magnetic flux which links with the

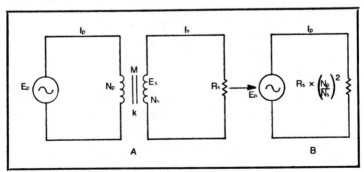

Fig. 12-5. Power transformer and its equivalent circuit.

secondary coil and induces the secondary voltage, E_s. With zero flux leakage, there are the same volts per turn associated with both the primary and secondary coils and therefore:

$$\text{Turns ratio} = \frac{N_p}{N_s} = \frac{E_p}{E_s}$$

If N_s is greater than N_p, E_s is greater than E_p and the transformer is referred to as "step-up."

Likewise, if N_s is less than N_p, E_s is less than E_p and the transformer is "step-down". The terms "step-up" and "step-down" normally refer to the voltage and not to the current. Note that if the primary and secondary coils are wound in the same sense, E_p, E_s are 180° out of phase.

The ideal transformer has zero power losses, and therefore the power input to the primary circuit equals the power output to the primary circuit equals the power output from the secondary circuit.

$$\text{Therefore: } E_p I_p = E_s I_s \text{ and: } \frac{E_p}{E_s} = \frac{I_s}{I_p} = \frac{N_p}{N_s}$$

This means that a step-up of the voltage level from the primary to the secondary circuit is accompanied by a corresponding reduction in the current level. The above equations may be arranged as:

$$E_p = \frac{E_s I_s}{I_p} = \frac{E_s N_p}{N_s}, \quad I_p = \frac{E_s I_s}{E_p} = \frac{I_s N_s}{N_p}$$

$$E_s = \frac{E_p I_p}{I_s} = \frac{E_p N_s}{N_p}, \quad I_s = \frac{E_p I_p}{E_s} = \frac{I_p N_p}{N_s}$$

$$N_p = \frac{E_p N_s}{E_s} = \frac{I_s N_s}{I_p}, \quad N_s = \frac{E_s N_p}{E_p} = \frac{I_p N_p}{I_s}$$

TRANSFORMER EFFICIENCY

The practical power transformer has the following losses:

1. The copper loss, which is the power dissipated in the resistances of the primary and secondary windings.

2. The iron loss, which is dissipated in the core. This may be subdivided into (a) the eddy-current loss, which is caused by the flux cutting the metallic core, and which may be reduced by

laminating the core; and (b) the hysteresis loss, which is the result of rapidly magnetizing, demagnetizing, and remagnetizing the core during the cycle of primary current.

These losses are taken into account by the efficiency percentage factor, η (Eta), which is defined by:

$$\eta = \frac{\text{power output from the secondary circuit}}{\text{power input to the primary circuit}} \times 100\%$$

$$= \frac{E_s I_s}{E_p I_p} \times 100\%$$

or the secondary power output is:

$$E_s I_s = E_p I_p \times \eta / 100$$

Then: $E_s = \dfrac{E_p I_p \eta}{I_s \times 100}$, $I_s = \dfrac{E_p I_p \eta}{E_s \times 100}$

and $E_p = \dfrac{E_s I_s \times 100}{\eta I_s}$, $I_p = \dfrac{E_s I_s \times 100}{\eta E_p}$

For the power losses in the transformer:

P_{loss} = primary power − secondary power

$$= \frac{100 - \eta}{100} \times \text{primary power}$$

$$= \frac{100 - \eta}{\eta} \times \text{secondary power}$$

For power transformers, the value of η normally exceeds 90%.

Note that if a steady direct current flows in the primary of a power transformer, the linkage flux will be constant in magnitude and direction, and the voltage induced in the secondary coil will be zero. However, if the steady DC voltage applied to the primary is mechanically "chopped" to produce a square wave, the transformer will respond to such an input and an alternating voltage (not, however, a simple sine wave) will be induced in the secondary; this secondary voltage may be rectified to produce a final DC output voltage which is larger than that of the DC input voltage. This is the principle used in the synchronous and nonsynchronous vibrator circuits.

REFLECTED RESISTANCE

If the secondary is loaded with a resistance R_s, then $R_s = E_s/I_s$. Since:

$$E_p = \frac{E_s N_p}{N_s} \text{ and } I_p = \frac{I_s N_s}{N_p}$$

Then:

$$\frac{E_p}{I_p} = \frac{E_s N_p/N_s}{I_s N_s/N_p} = \frac{E_s}{I_s} \times \left(\frac{N_p}{N_s}\right)^2 = R_s \times \left(\frac{N_p}{N_s}\right)^2$$

The equation $\quad \dfrac{E_p}{I_p} = R_s \times \left(\dfrac{N_p}{N_s}\right)^2$

may be represented by the equivalent circuit shown in Fig. 12-5 B. The expression $R_s \times (N_p/N_s)^2$ is the effective resistive load presented to the primary source and is referred to as the value of resistance reflected from the secondary circuit into the primary circuit due to the introduction of the secondary load R_s. If the turns ratio of the transformer is chosen so that the value of the reflected resistance, $R_s \times (N_p/N_s)^2$, is equal to the internal resistance of the primary source, the secondary load is then matched to the primary source for maximum power transfer to the secondary load.

If R_p is the resistance associated with the primary source, the condition for matching is:

$$R_p = R_s \left(\frac{N_p}{N_s}\right)^2 \text{ and } R_s = R_p \left(\frac{N_s}{N_p}\right)^2$$

or $\qquad \dfrac{R_p}{R_s} = \left(\dfrac{N_p}{N_s}\right)^2 \text{ and } \dfrac{N_p}{N_s} = \sqrt{\dfrac{R_p}{R_s}}$

Example 12-4

In Fig. 12-5 A, $E_p = 110$ V, 60 Hz, $N_p = 1500$ turns, $N_s = 6000$ turns, $R_s = 180\,\Omega$. Assuming a coupling factor of unity and that the transformer is 100% efficient, calculate the values of E_s, I_s, I_p, primary power, secondary power and the reflected resistance.

Solution

$$\text{Turns ratio} = \frac{N_p}{N_s} = \frac{1500}{6000} = 1:4.$$

$$\text{Secondary voltage, } E_s = E_p \times \frac{N_s}{N_p} = 110 \text{ V} \times 4$$
$$= 440 \text{ V}.$$

$$\text{Secondary current, } E_s = \frac{E_s}{R_s} = \frac{440 \text{ V}}{180 \text{ }\Omega} = 2.444 \text{ A}.$$

$$\text{Primary current, } I_p = I_s \times \frac{N_s}{N_p} = 2.444 \times 4 = 9.776 \text{ A}.$$

$$\text{Primary power} = \text{Secondary power} = E_p \times I_p = 110 \text{ V} \times 9.776$$
$$= 1075 \text{ W}.$$
$$\text{Check: Secondary power} = E_s \times I_s = 440 \text{ V} \times 2.444 \text{ A}$$
$$= 1075 \text{ W}.$$

$$\text{Reflected resistance} = \frac{E_p}{I_p} = \frac{110 \text{ V}}{9.776 \text{ A}} = 11.25 \text{ }\Omega.$$

$$\text{Check: Reflected resistance} = R_s \times \left(\frac{N_p}{N_s}\right)^2 = \frac{180}{16}$$

$$= 11.25 \text{ }\Omega.$$

Example 12-5

In Fig. 12-5 A, E_p = 220 V, 60 Hz, I_p = 1.2 A, E_s = 55 V, R_s = 12 Ω. Calculate the values of the primary power, secondary power and transformer efficiency.

Solution

$$\text{Primary power} = E_p \times I_p = 220 \text{ V} \times 1.2 \text{ A} = 264 \text{ W}.$$

$$\text{Secondary power} = \frac{E_s^2}{R_s} = \frac{(55 \text{ V})^2}{12 \text{ }\Omega} = 252 \text{ W}.$$

$$\text{Transformer power loss} = 264 - 252 = 12 \text{ W}.$$

$$\text{Transformer efficiency} = \frac{252}{264} \times 100 = 95\%.$$

CHAPTER SUMMARY

☐ *Mutually Coupled Circuits.*

Coupling factor, $k = \dfrac{M}{\sqrt{L_1 L_2}}$, Mutual inductance, $M = k \times \sqrt{L_1 L_2}$

☐ *Mutually Coupled Coils*

Series-aiding, $L_T^+ = L_1 + L_2 + 2M$. Series-opposing, $L_T^- = L_1 + L_2 - 2M$

Parallel-aiding, $L_T^+ = \dfrac{L_1 L_2 - M^2}{L_1 + L_2 + 2M}$. Parallel-opposing, L_T^-

$$= \dfrac{L_1 L_2 - M^2}{L_1 + L_2 - 2M}$$

☐ *Power Transformer.*

100% Efficiency, coupling factor = 1.
Turns Ratio,

$$\frac{N_p}{N_s} = \frac{E_p}{E_s} = \frac{I_s}{I_p}.$$

Primary Power, $E_p I_p$ = Secondary power, $E_s I_s$.
Resistance reflected into the primary circuit

$$= R_s \times \left(\frac{N_p}{N_s} \right)^2.$$

☐ *Practical Transformer.*

Coupling factor less than 1.

Transformer efficiency, $\eta = \dfrac{\text{secondary power}}{\text{primary power}} \times 100\%$

$$= \frac{E_s I_s}{E_p I_p} \times 100\%.$$

Complex Algebra

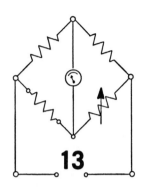

13

In previous chapters the alternating voltage or current has been represented either by a sine waveform or a phasor or a trigonometric expression. While these representations are adequate for simple series and parallel arrangements, they are too cumbersome for the analysis of more complicated circuits which require the use of the network theorems. What is needed is a form of algebra which can be applied directly to the solution of AC circuits. Such an algebra must be capable of taking into account the circuit's phase relationships by distinguishing between the resistive and the reactive elements. This is achieved in complex algebra by the introduction of the *operator j*.

INTRODUCTION TO COMPLEX ALGEBRA AND TO THE OPERATOR j

A phasor when multiplied by the operator j is rotated through $90°$ or $\frac{\pi}{2}$ radians in the positive or counterclockwise direction but the magnitude of the phasor is unchanged.

Referring to Fig. 13-1, OP represents a phasor in the horizontal reference position. From the definition of operator j:

Phasor $OP_1 = j \times$ Phasor OP
Phasor $OP_2 = j \times$ Phasor $OP_1 = j^2 \times$ Phasor OP
Phasor $OP_3 = j \times$ Phasor $OP_2 = j^2 \times$ Phasor OP_1
$\qquad\qquad\qquad\qquad\quad = j^3 \times$ Phasor OP.

Since phasors OP and OP_2 are $180°$ apart,
Phasor $OP_2 = -$ Phasor OP
but Phasor $OP_2 = j^2 \times$ Phasor OP
Therefore: $j^2 = -1$.

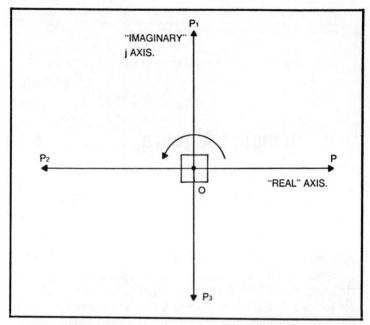

Fig. 13-1. Operator j.

Since the square of a real positive or negative number is always positive, j cannot be evaluated in terms of normal numbers and is therefore known as an "imaginary" quantity. As shown in Fig. 13-1 a phasor when multiplies by j lies along the vertical direction which may therefore referred to as the j or "imaginary" axis. By contrast the horizontal direction will be referred to as the "real" axis.

Since $j^3 = j^2 \times j = -1 \times j = -j$, the multiplication of a phasor by $-j$ will cause the phasor to be rotated throught 90° or $\dfrac{\pi}{2}$ radians in the negative or clockwise direction, leaving the magnitude of the phasor unchanged. Note that since $j^2 = -1$, $-j$ and $+j$ are reciprocals.

RECTANGULAR AND POLAR NOTATION

Figure 13-2 represents a phasor diagram in which Phasor OP = Phasor r, Phasor OM = Phasor × and Phasor OQ = Phasor z.

Then Phasor ON = Phasor jx and since Phasor OQ = Phasor ON + Phasor OP, then z = r + jx.

244

This is known as the rectangular notation since the phasors r and jx are 90° apart. Since phasor z is specified in terms of *two* phasors whose order is important ($z = 2 + j^3$ is not the same as $z = 3 + j^2$), this representation of a phasor is referred to as complex algebra.

The polar method of denoting a phasor is in terms of its magnitude and its direction. The magnitude of z is represented by the length of the line OQ and the direction is measured by the angle, ϕ, between OQ and the horizontal reference line. In this polar notation,

$$z = \underline{/Z \ \phi}$$

where Z is the magnitude of the phasor z; the value of ϕ will lie between 0° and 180°.

If the phasor lies in the lower two quadrants, the angle is negative and then $z = Z \underline{/-\phi}$ or $Z \overline{\phi}$ where $Z \underline{/-\phi} = Z \overline{\phi}$. $Z \overline{\phi}$ should not be confused with $-\underline{/Z} \ \phi$ which equals $Z \underline{/\phi + \pi}$. Conversions between rectangular and polar notations:

If R and X are the magnitudes of the phasors r and x,

$$Z = + \sqrt{R^2 + X^2} \text{ and } \phi = \text{Inv. tan.} \ \frac{X}{R}$$

Note that Z is always considered to be positive.

Also R = Z Cos ϕ, X = Z Sin ϕ.

As an example, consider a resistance of 3 Ω in series with an inductive reactance of 4 Ω. Referring to chapter ten, the phasor

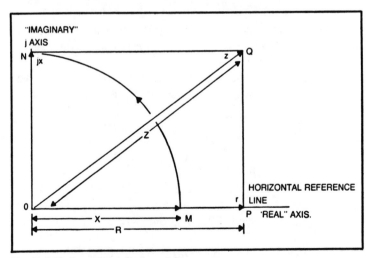

Fig. 13-2. Rectangular and polar notations.

diagram would show the resistance phasor to be along the horizontal reference line while the reactance phasor would be vertical in the direction of the j axis. The resultant of the resistance and the reactance phasors is the impedance phasor, z. In rectangular notation

$$z = 3 + j4.$$

Impedance magnitude, $Z = +\sqrt{3^2 + 4^2} = 5 \, \Omega$

Phase angle, $\phi = $ Inv. tan. $\dfrac{4}{3} = +53°8'$

Then $z = 3 + j4 = 5\underline{/53°8'}$

and $z = 5\underline{/53°8'} = 5 \cos 53°8' + j\,5 \sin 53°8'$
$$= 3 + j4.$$

Many electronic calculators have the capability of rapid conversion between rectangular and polar notations. You should familiarize yourself with the methods used in your own calculator.

Example 13-1

Convert (a) $z = 2 - j3$, $z = -5 + j2$ and $z = -4 - j7$ into polar notation and (b) $z = 6\ \underline{/132°}$, $z = 8\ \underline{/-72°}$, $z = 12\ \underline{/-117°}$ into rectangular notation.

Solution

(a) $z = 2 - j3 = +\sqrt{2^2 + 3^2}\underline{/\text{Inv. Tan.}(-3/2)}$
$$= 3.61\ \underline{-/56.3°}.$$

Caution should be exercised in the determination of the angle. It is advisable to use a rough sketch to establish the quadrant in which the resultant phasor lies.

It should also be remembered that the angle derived mathematically is always measured with respect to the horizontal line.

$z = -5 + j2 = +\sqrt{5^2 + 2^2}\underline{/\text{Inv. Tan.}(-5/2)}$
$$= 5.39\ \underline{/111.2°}.$$
$z = -4 - j7 = +\sqrt{4^2 + 7^2}\underline{/\text{Inv.Tan.}(-7/4)}$
$$= 8.06\ \underline{-/119.7}.$$

(b) $z = 6\ \underline{/132°} = 6 \cos 132° + j6 \sin 132°$
$$= -4.01 + j\,4.46.$$
$z = 8 - \underline{/72°} = 8 \cos(-72°) + j8 \sin(-72°)$
$$= 2.47 - j7.61.$$
$z = 12 - \underline{/117°} = 12 \cos(-117°) + j12 \sin(-117°)$
$$= 5.45 - j\,10.7.$$

RULES OF COMPLEX ALGEBRA

1. *Equating real and imaginary parts.*
Let $z_1 = R_1 + jX_1$, and $z_2 = R_2 + jX_2$.
If $z_1 = z_2$, $R_1 + jX_1 = R_2 + jX_2$.
Then $R_1 = R_2$ and $X_1 = X_2$.

2. *Addition and Subtraction of Phasors.*
If $z_1 = R_1 + jX_1$ and $z_2 = R_2 + jX_2$, then
$$z_1 + z_2 = (R_1 + jX_1) + (R_2 + jX_2)$$
$$= (R_1 + R_2) + j(X_1 + X_2)$$
and $z_1 - z_2 = (R_1 - R_2) + j(X_1 - X_1)$.

Rectangular rather than polar notation is used when adding or subtracting phasors.

3. *Multiplication of Phasors.*

If $z_1 = R_1 + jX_1$ and $z_2 = R_2 + jX_2$, then
$$z_1 z_2 = (R + jX)(R + jX_5)$$
$$= R_1 R_2 + jR_m X_2 + jR_2 X_1 + j^2 X_1 X_2$$
$$= (R_1 R_2 - X_1 X_2) + j(R_1 X_2 + R_2 X_1)$$
since $j_2 = -1$.

Using polar notation, $z_1 = Z_1 \underline{/\phi 1}$
and $z_2 = Z_2 \underline{/\phi 2.}$
Then $z_1 z_2 = Z_1(\cos\phi_1 + j\sin\phi_1) \times Z_2(\cos\phi_2 + j\sin\phi_2)$
$$= Z_1 Z_2 [(\cos\phi_1 \cos\phi_2 - \sin\phi_1 \sin\phi_2)$$
$$+ j(\sin\phi_1 \cos\phi_2 + \cos\text{-} \phi \sin\phi_1)]$$
$$= Z_1 Z_2 [\cos(\phi_1 + \phi_2) + j\sin(\phi_1 + \phi_2)]$$
$$= Z_1 Z_2 \underline{/\phi_1 + \phi_2}$$

When multiplying phasors, the magnitudes are multiplied but the angles are added. It is preferable to use polar nitation when multiplying or dividing phasors.
If $z = Z\underline{/\phi}$,
$$z^2 = Z\underline{/\phi} \times Z\underline{/\phi} = Z^2 \underline{/2\phi.}$$
Therefore in squaring a phasor, the magnitude must be squared but the angle is doubled. It follows that in taking the square root of a phasor, you must obtain the square root of the magnitude but the angle is halved.

$$\sqrt{z} = \sqrt{Z}\underline{/\phi} = \sqrt{Z}\underline{/\phi/2} = Z^{\frac{1}{2}}\underline{/\phi/2}.$$

4. *Division of Phasors.*

If $z_1 = R_1 + jX_1$, $z_2 = R_2 + jX_2$.

$$\frac{z_1}{z_2} = \frac{R_1 + jX_1}{R_2 + jX_2}$$

In order to separate our the real and imaginary parts of z_1/z_2, it is necessary to eliminate j from the denominator. This is done by "rationalization" which means multiplying both numerator and denominator by the "conjugate" of the denominator. The conjugate of a phasor is that phasor which has the same magnitude but whose angle is opposite in sign although equal in magnitude. Therefore the conjugate of $Z\underline{/\phi}$ is $Z\underline{/\phi}$ and the conjugate of $R + jX$ is $R - jX$.

Then:
$$\frac{z_1}{z_2} = \frac{(R_1 + jX_1)(R_2 - jX_2)}{(R_2 + jX_2)(R_2 - jX_2)}$$

$$= \frac{(R_1R_2 + X_1X_2)}{R_2^2 + X_2^2} + \frac{j(X_1R_2 - X_2R_1)}{R_2^2 + X_2^2}$$

In polar form $z_1 = Z_1\underline{/\phi_1}$, $= Z_2\underline{/\phi_2}$ so that:

$$\frac{z_1}{z_2} = \frac{Z_1(\cos\phi_1 + j\sin\phi_1)}{Z_2(\cos\phi_2 + j\sin\phi_2)}$$

$$= \frac{Z_1}{Z_2} \times \left[\frac{(\cos\phi_h + j\sin\phi_1)(\cos\phi_2 - j\sin\phi_2)}{(\cos\phi_2 + j\sin\phi_2)(\cos\phi_2 - j\sin\phi_2)}\right]$$

$$= \frac{Z_1}{Z_2} \times \left[\frac{(\cos\phi_1 \cos_2 + \sin\phi_1 \sin\phi_2) + j(\sin\phi_1 \cos\phi_2 - \cos\phi_1}{\cos^2\phi_2 + \sin\phi_2}\right]$$

$$= \frac{Z_1}{Z_2} \times \left[\frac{\cos(\phi_1 - \phi_2) + j\sin(\phi_1 - \phi_2)}{1}\right]$$

$$= Z_1/Z_2 \underline{/\phi_1 - \phi_2}.$$

When dividing phasors, the magnitudes are divided but the angles are subtracted. Note that in the special case of the reciprocal of $z = Z\underline{/\phi°}$,

$$\frac{1}{z} = \frac{1}{Z} \frac{\big/ 0°}{\big/ \phi°} = \frac{1}{Z} \; \underline{\big/ -\phi°}$$

See Figs. 13-3A, 13-3B and 13-3C.

Example 13-2

If $z_1 = 3 + j5$ and $z_2 = 4 - j7$

Solution

$z_1 = 3 + j5 = \sqrt{3^2 + 5^2} \underline{\big/ \text{Inv. Tan. } 5/3} = 5.83 \; \underline{\big/ 59.0°}.$
$z_2 = 4 - j7 = \sqrt{4^2 + 7^2} \underline{\big/ \text{Inv. Tan. } (-7/4)} = 8.06 \; \underline{\big/ -60.3°}.$

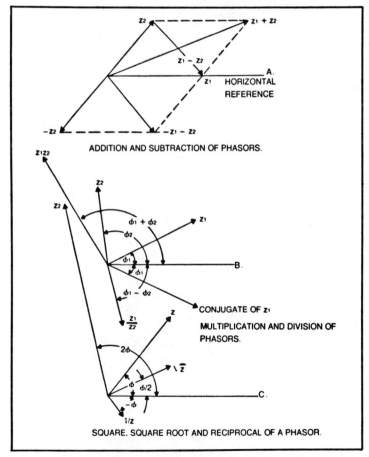

Fig. 13-3. Phasor illustrations of the rules of complex algebra.

$$z_1 + z_2 = (3 + j5) + (4 - j5) = 7 - j2$$
$$= \sqrt{7^2 + 2^2}\ \underline{/\text{Inv. Tan. } (- 2/7)}$$
$$= 7.28\ \underline{/-16.0°}.$$

$$z_1 - z_2 = (3 + j5) - (4 - j7) = -1 + j12$$
$$= \sqrt{1^2 + 12^2}\ \underline{/\text{Inv. Tan. } 12/1}$$
$$= 12.04\ \underline{/94.8°}.$$

$$z_1 z_2 = 5.83\ \underline{/59.0°} \times 8.06\ \underline{/-60.3°} = 50.0\ \underline{/-1.3°}.$$

$$z_1/z_2 = 5.83\ \underline{/59.°}/8.06\ \underline{/-60.3°} = 0.723\ \underline{/119.3°}.$$

$$z_1^{\ 2} = 5.8^2\ \underline{/2 \times 59.0°} = 33.64\ \underline{/118°}.$$

$$\sqrt{z_2} = \sqrt{8.06}\ \underline{/-60.3/2°} = 2.84\ \underline{/-30.15°}.$$

$$z_1 = z_1^{-1} = \frac{1\ \underline{/0°}}{5.83\ \underline{/59.0°}} = 0.172\ \underline{/-59.0°}$$

ANALYSIS OF AC CIRCUITS

Resistance

$$Z = \frac{E_{rms}}{I_{rms}} = R, \phi = 0°.$$

$$\text{Phasor } z = R\ \underline{/0°} = R + j0.$$

Inductance

$$Z = \frac{E_{rms}}{I_{rms}} = \omega L = 2\pi f L = X_L, \phi = + 90°.$$

$$\text{Phasor } z = \omega L \underline{\left/\frac{\pi}{2}\right.} = 0 + j\omega L.$$

Capacitance

$$Z = \frac{E_{rms}}{I_{rms}} = \frac{1}{\omega C} = \frac{1}{2\pi f C} = X_C, \phi = - 90°.$$

$$\text{Phasor } z = \frac{1}{\omega C}\ \underline{-\left/\frac{\pi}{2}\right.} = 0 - \frac{j}{\omega C} = 0 + \frac{1}{j\omega C}$$

Resistance, Inductance and Capacitance in Series

$$\text{Phasor } z = R + j\omega L - \frac{j}{\omega C}$$

$$\frac{E_{rms}}{I_{rms}} = Z = \sqrt{R^2 + \left(\omega L - \frac{1}{\omega C}\right)^2}$$

Phase angle, $\phi = $ Inv. Tan. $\left(\dfrac{\omega L - \dfrac{1}{\omega C}}{R} \right)$

Resistance, Inductance and Capacitance in Parallel

$$\frac{1}{z} = \frac{1}{R} + \frac{1}{j\omega L} + \frac{1}{-j/\omega C} = \frac{1}{R} + \frac{1}{j\omega L} + \frac{j}{\omega C}$$

or $y = G - jB_L + jB_C$

Then: $Y = \dfrac{1}{Z} = \sqrt{G^2 + (B_C - B_L)^2}$

and $\phi = $ Inv. Tan. $\left(\dfrac{B_C - B_L}{G} \right)$

The use of operator j in the analysis of more complex AC circuits is illustrated in the following examples.

Example 13-3

In the series-parallel circuit of Fig. 13-4, express the total impedance phasor and the supply current in their polar forms.

Solution

$z_1 = 3 + j5$ and $z_2 = 4 - j2$.

Using the product-over-sum formula for the total impedance phasor,

$$z_T = \frac{z_1 z_2}{z_1 + z_2} = \frac{(3 + j5)(4 - j2)}{(3 + j5) + (4 - j2)} = \frac{(3 + j5)(4 - j2)}{(7 + j3)}$$

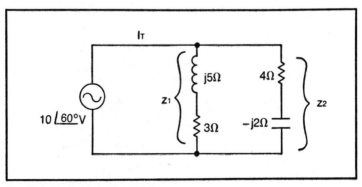

Fig. 13-4. Circuit for example 13-3.

Since we are now faced with multiplication and division of phasors, it is more convenient to convert to the polar form. Then:

$$z_T = \frac{\sqrt{3^2 + 5^2}\,\underline{/\text{Inv. Tan. } 5/3} \times \sqrt{4^2 + 2^2}\,\underline{/\text{Inv. } (-2/4)}}{\sqrt{7^2 + 3^2}\,\underline{/\text{Inv. Tan. } 3/7}}$$

$$= \frac{5.83\,\underline{/59.04^\circ} \times 4.47\,\underline{/-26.57^\circ}}{7.62\,\underline{/23.20^\circ}}$$

$$= \frac{5.83 \times 4.47}{7.62}\,\underline{/59.04^\circ - 26.57^\circ - 23.20^\circ}$$

$$= 3.42\,\underline{/9.27^\circ}\,\Omega$$

Supply current:

$$I_T = \frac{10\,\underline{/60^\circ}}{3.42\,\underline{/9.27^\circ}} = 2.92\,\underline{/50.73^\circ}\,\text{A}.$$

Example 13-4

In Fig. 13-5A derive the Thévenin equivalent circuit between the terminals X, Y and hence obtain the load current in its polar form.

Solution

Step 1. Remove the load and calculate the open-circuit voltage between X and Y. Using the voltage division rule:

$$e_{TH} = 10\,\underline{/30^\circ} \times \frac{j4}{(1 - j3) + j4} = 10\,\underline{/30^\circ} \times \frac{4\,\underline{/90^\circ}}{1 + j}$$

$$= \frac{10\,\underline{/30^\circ} \times 4\,\underline{/90^\circ}}{\sqrt{2}\,\underline{/45^\circ}}$$

$$= 28.3\,\underline{/75^\circ}\,\text{V}.$$

Step 2. Place a short circuit across the voltage source and calculate the value of the impedance phasor between X and Y.

$$z_{TH} = -j2 + \frac{j4 \times (1 - j3)}{j4 + (1 - j3)} = -2j + \frac{j4 + 12}{1 + j}$$

$$= -2j + \frac{(j4 + 12)(1 - j)}{2}$$

$$= -2j + 8 - j4$$

$$= 8 - j6\,\Omega.$$

Step 3. In the Thévenin equivalent circuit of Fig. 13-4B, replace the load. Then the load current is

$$i_L = \frac{e_{TH}}{z_{TH} + R_L} = \frac{28.3 \underline{/75°}}{8 - j6 - 5} = \frac{28.3 \underline{/75°}}{13 - j6}$$

$$= \frac{28.3 \underline{/75°}}{14.3 \underline{/-24.78°}}$$

$$= 1.98 \underline{/99.78°} \text{ A.}$$

Example 13-5

In Fig. 13-5A derive the Norton equivalent circuit between the terminals X, Y and hence obtain the load current in its polar form.

Solution

Step 1. Remove the load and replace it by a short circuit. The toal impedance then presented to the voltage source is:

$$z_T = 1 - j3 + \frac{j4 \times (- j2)}{j4 + (- j2)}$$

$$= 1 - j3 + \frac{8}{j2} = 1 - j3 - j4$$

$$= 1 - j7.$$

The current drawn from the source is

$$i_T = \frac{10 \underline{/30°}}{1 - j7}$$

and the short circuit current between the terminals, X, Y is:

$$i_N = \frac{10 \underline{/30°}}{1 - j7} \times \frac{j4}{j4 + (- j2)}$$

$$= \frac{20 \underline{/30°}}{7.07 \underline{/-81.87°}} = 2.83 \underline{/111.87°} \text{ A.}$$

Step 2. The Norton impedance, z_N, is the same as the Thévenin impedance, z_{TH}, and therefore $z_N = 8 - j6$. See Fig. 13-5B.

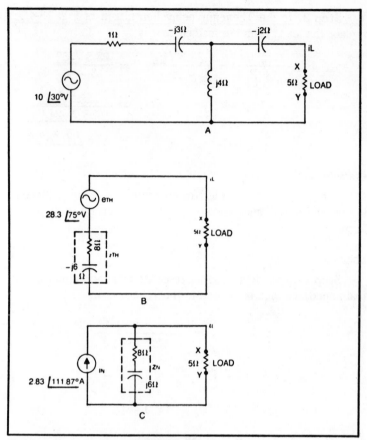

Fig. 13-5. Circuit for example 13-4 and 13-5.

Step 3. Remove the short circuit and replace the load in the Norton equivalent circuit of Fig. 13-5C. By the current division rule:

$$i_L = i_N \times \frac{8 - 6j}{8 - 6j + 5} = 2.83 \underline{/111.87°} \times \frac{8 - 6j}{13 - 6j}$$

$$= \frac{2.83 \underline{/111.87°} \times 10 \underline{/-36.87°}}{14.32 \underline{/-24.78°}}$$

$$= 1.98 \underline{/99.78°} \text{ A.}$$

Example 13-6

In the circuit of Fig. 13-6A express the current i_L in its polar form by the use of the superposition theorem.

Solution

To use the superposition theorem, place a short circuit across the 10 $\angle -45°$ V source (Fig. 13-6B). The total impedance then presented to the 15 $\angle -60°$ V source is:

$$z_{IT} = -j4 + \frac{3 \times j5}{3 + j5} = -j4 + \frac{j15\,(3 - j5)}{(3 + j5)\,(3 - j5)}$$

$$= -j4 + \frac{75 + j45}{34}$$

$$= -j4 + 2.21 + j\,1.32$$
$$= 2.21 - j\,2.68 \ \Omega.$$

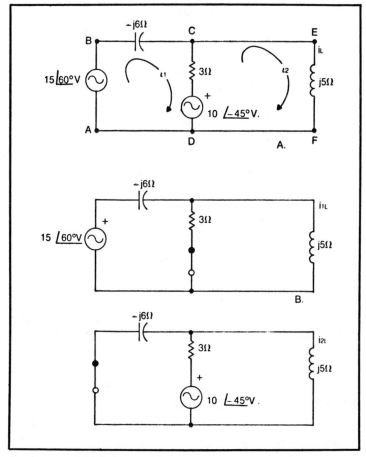

Fig. 13-6. Circuits for example 13-6.

Then:
$$i_{IT} = \frac{15 \angle 60°}{2.21 - j\,2.28}$$

and
$$i_{IL} = \frac{15 \angle 60°}{2.21 - j\,2.68} \times \frac{3}{3 + j5}$$

$$= \frac{45 \angle 60°}{3.47 \angle -50° \times 5.83 \angle 59.04°}$$

$$= 2.22 \angle 51.46° \text{ A}$$

The next step is to short out the $15 \angle 60°$ V source (Fig. 13-6C). The total impedance presented to the $10 \angle -45°$ V source is:

$$z_{2T} = 3 + \frac{(-j4) \times j5}{j5 - j4} = 3 + \frac{20}{j}$$

$$= 3 - j20.$$

Then: $i_{2T} = \dfrac{10 \angle -45°}{3 - j20}$

and $i_{2L} = \dfrac{10 \angle -45°}{3 - j20} \times \dfrac{(-j4)}{(-j4 + j5)}$

$$= \frac{-40 \angle -45°}{3 - j20} = \frac{-40 \angle -45°}{20.22 \angle -81.47°}$$

$$= -1.98 \angle 36.47° \text{ A}.$$

Total $i_L = i_{1L} + i_{2L} = 2.22 \angle 51.46° - 1.98 \angle 36.47°$

$$= 1.38 + j\,1.736 - 1.592 - j\,1.177$$

$$= -0.212 - j\,0.559$$

$$= 0.6 + \angle 110° \text{ A, rounded off.}$$

For comparison, example 13-16 will be solved by the method of mesh analysis.

MESH ABCD

$15 \angle 60° - 10 \angle -45° = i_1(3 - j4) - i_2\,3$

$7.5 + j\,12.99 - 7.07 + j\,7.07 = i_1(3 - j4) - i_2\,3$

$0.43 + j\,20.06 = i_1(3 - j4) - i_2\,3.$

MESH CEFD

$10 \angle -45° = (i_2 - i_2)\,3 + i_2 j5$

$7.07 - j\,7.07 = -i_1\,3 + i_2\,(3 + j5)$

Multiplying the two preceeding equations by $(3 - j4)$:

$1.29 + j\,60.18 = i_1\,(9 - j\,12) - i_2 9.$

$21.21 - j49.49 - 28.28 = -i_1\,(9 - j\,12)$
$$+ i_2\,(3 - j4)\,(3 + j5)$$

Adding the two preceeding equations:

$- 5.98 + j\,10.69 = i_2\,(20 + j3).$

$$\text{Load current, } i_L + i_2 = \frac{-5.98 + j\,10.69}{20 + j3}$$

$$= \frac{12.15\ \underline{/118.4°}}{20.22\ \underline{/8.53°}}$$

$$= 0.6\ \underline{/110°}\text{ A, rounded off.}$$

CHAPTER SUMMARY

☐ *Rectangular to Polar Notation.*

$z = R + jX = Z\ \underline{/\phi}$ where $Z = \sqrt{R^2 + X^2}$ and $\phi = \text{Inv. Tan.}\dfrac{X}{R}$

☐ *Polar to Rectangular Notation.*

$z = Z\ \underline{/\phi} = Z\cos\phi + jZ\sin\phi = R + jX$ where $R = Z\cos\phi$ and $X = Z\sin\phi$.

☐ *Rules of Complex Algebra.*

$z_1 = Z_1\ \underline{/\phi_1} = R_1 + j\,X_1$, $z_2 = Z_2\ \underline{/\phi_2} = R_2 + j\,X_2$

Equating "real" and "imaginary" parts.

If $R_1 + j\,X_1 = R_2 + j\,X_2$, $R_1 = R_2$ and $X_1 = X_2$.

Addition: $z_1 + z_2 = (R_1 + R_2) + j\,(X_1 + X_2)$.

Subtraction: $z_1 - z_2 = (R_1 - R_2) + j\,(X_1 - X_2)$.

Multiplication: $z_1\,z_2 = Z_1\,Z_2\ \underline{/\phi_1 + \phi_2}$

Square: $z_1^2 = Z_1^2\ \underline{/2\phi_1}$

Square Root: $\sqrt{z_1} = \sqrt{Z_1}\ \underline{/\phi_{1/2}}$

Division:

$$\frac{z_1}{z_2} = \frac{Z_1}{Z_2} \underline{/\phi_1 - \phi_2}$$

Reciprocal:

$$\frac{1}{z_1} = \frac{1}{Z_1} \underline{/-\phi_1}$$

Conjugates: The conjugate of $z = Z \underline{/\phi} = R + jX$ is $z^1 = Z \underline{/-\phi} = R - jX$.

Rationalization: To rationalize

$$\frac{A + jB}{C + jD},$$

multiply numerator and denominator by the denominator's conjugate, $C - jD$.

Electrical Measurements

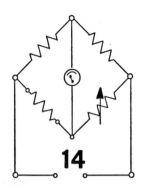

14

The most common type of indicating instrument is fitted with a pointer which shows on a scale the value of the quantity being measured. Such an instrument has a moving system which is carried by a hardened steel spindle with its ends tapered and highly polished. The tapered ends form pivots which rest in hollow-ground sapphire bearings set in steel screws.

INTRODUCTION TO INDICATING INSTRUMENTS

These indicating instruments all possess three essential features:

1. A deflecting device which enables a mechanical force to be exerted by the voltage, current or power being measured.

2. A controlling device which ensures that the amount of the deflection is dependent on the magnitude of the measured quantity.

3. A damping device which prevents oscillation of the moving system and enables the final position to be reached quickly.

The various devices used depend on the particular type of instrument and its purpose.

MOVING-COIL (D'ARSONVAL) METER MOVEMENT

Referring to Fig. 14-1 the deflecting device of the moving-coil meter movement consists of a rectangular coil, C, which is made of thin insulated copper wire wound on a light aluminum former. This frame is carried by a spindle which pivots in jewelled bearings. Current is led into and out of the coil by the spiral hairsprings, H, which are the controlling device and therefore provide the restoring torque. The coil is free to move in the gaps between the permanent magnet pole-pieces, P, and a soft iron cylinder, A, which is normally carried by a non-magnetic bridge attached to P. The purpose of the soft iron cylinder is

1. to concentrate the magnetic flux by reducing the amount of the air-gap.

2. to produce a radial magnetic field with a uniform flux density. This is also achieved by special shaping of the pole pieces, P.

Let the coil consist of N turns, each with a length of L meters and a width of d meters.

In a radial field the torque, T, exerted on the coil is:

$$T = BINLd = BANI \text{ newton-meters}$$

where:
    I = Current flowing through coil in amperes.
    B = Flux density in teslas.
    A = Area of coil in square meters.

The restoring torque, T', provided by the hairsprings, is directly proportional to the amount of the deflection, 0°. Therefore T' = k0 where k is a constant, and in the final equilibrium position:

$$T = T'$$

$$BANI = k\theta$$

$$I = \frac{k\theta}{BAN} = K\theta$$

where K is the constant of proportionality for the entire meter movement.

The deflection is directly proportional to the current so that the scale is linear with the divisions evenly spaced.

As the coil swings towards the equilibrium position, it gains momentum which could create unwanted oscillation. The required damping is introduced by winding the coil on the aluminum frame. As the coil rotates, eddy currents are induced into the frame and these absorb the kinetic energy of the system. When the coil reaches its equilibrium position, it has no energy to swing further and since the deflecting and restoring torques are balanced, efficient damping is obtained.

The sensitivity of the moving coil meter movement will be determined by the amount of current required to provide full-scale deflection (The lower the value of current, the greater the movement's sensitivity). High sensitivity will be achieved by using a strong permanent magnet to produce a large flux density; in addition many turns of fine copper wire will be wound on a light aluminum frame and then attached to delicate hairsprings. The full scale deflection current for such a meter movement would be 20μA

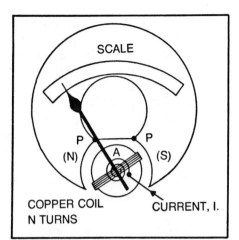

Fig. 14-1. Moving coil meter movement.

SCALE

P (N)   P (S)

A

COPPER COIL
N TURNS

CURRENT, I.

or less. By using fewer turns of the larger wire size together with stronger hairsprings, the full scale deflection current can be increased to a few mA; the limitation on the current then becomes the physical size of the meter movement.

Since the sensitivity and the full scale deflection current are inversely related, the sensitivity is directly measured in terms of the current's reciprocal. With current equivalent to volts divided by ohms, the units of sensitivity will be ohms per volt. For example if the meter movements fullscale deflection current is 50 μA, the

sensitivity is $1/50 \ \mu A = \dfrac{1}{50 \times 10^{-6}} = 20000$ ohms per volt.

Therefore: Sensitivity (ohms per volt) =

$$\dfrac{1}{\text{Full scale deflection current (amperes)}}$$

and

Full scale deflection current = $\dfrac{1}{\text{Sensitivity (ohms per volt)}}$
(amperes)

The moving coil meter movement is therefore basically a microammeter or milliammeter. Typically the resistance of the coil is of the orders of tens of ohms and therefore the meter movement can only be calibrated to measure millivolts.

The passage of current through the coil raises its temperature and alters its resistance. It is normal practice to add a swamp resistor in series with the coil. The swamp resistor has a negative temperature coefficient which counteracts the positive temperature coefficient of the copper coil. It is also used to bring the total resistance (coil resistance plus swamp resistance) to some suitable value such as 50 Ω. It must be remembered that in order to measure current at a particular point, the circuit must be broken and the meter inserted in the break. The total 50 Ω resistance of the meter movement will be in series with the remainder of the circuit and must therefore affect the reading to a certain effect. The amount of the error will depend on the value of the circuit's total resistance in comparison with the 50 Ω resistance of the meter movement.

The main advantages of the moving coil meter movement are:
1. uniform scale.
2. high sensitivity.
3. high degree of shielding from stray magnetic fields.
The principal disadvantages are:
1. Only suitable for DC measurement.
2. Relatively expensive when compared with the moving-iron instrument.

Example 14-1

The sensitivity of a moving coil meter movement is 1000 Ω/V. What is the value of the full-scale deflection current?

Solution

$$\text{The full-scale deflection current} = \frac{1}{1000 \ \Omega/V}$$

$$= \frac{1}{1000} \ A = 1 \ mA.$$

THE AMMETER

In order to convert the basic moving coil meter movement into an instrument capable of measuring hundreds of milliamperes or even amperes, a shunt resistor is connected across the series combination of the meter movement and the swamp resistor, (Fig. 14-2). Since the shunt resistor has a low value, most of the current to be measured will be diverted through R_{sh} and only a small fraction will pass through the meter movement to provide the required deflection.

Then: $I = I_m + I_{sh}$.
and $I_{sh} \times R_{sh} = I_m \times R_m$.

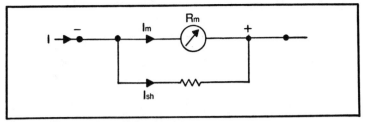

Fig. 14-2. The ammeter and the meter shunt.

$$R_{sh} = \frac{I_m \times R_m}{I_{sh}} = \frac{I_m \times R_m}{I - I_m}$$

where: I = current to be measured

I_m = current flowing through the meter movement

R_{sh} = shunt resistance

R_m = total meter movement resistance.

Figure 14-3 shows an instrument which has a number of shunts for different current ranges. However, there is a danger that when switching from range to range, the meter movement will be placed directly in the circuit without the protection of a shunt; this may be avoided by making the switch of a make-before-break type. Alternatively the instrument may use a Ayrton shunt, (Fig. 14-4), which ensures that the meter cannot be placed in the circuit without the presence of a shunt.

Because of its very low resistance the ammeter must never be placed directly across a voltage source. When measuring an unknown current the meter should initially be switched to the highest range. If the reading is then too small, the meter can be switched to a lower range; in this way the meter movement is protected from excessive current.

Example 14-2

In Fig. 14-3 the full scale deflection current of the meter movement is 1 mA and its resistance is 50 Ω. Calculate the values of the shunt resistors for the ranges of (a) 0 – 10 mA (b) 0 – 100 mA and (c) 0 – 1 A.

Solution

(a). The voltage across the meter movement = 1 mA × 50 Ω = 50 mV. For the full scale current of 10 mA, the shunt current is 10 mA – 1 mA = 9 mA. The value of the shunt resistor is

$$= \frac{50 \text{ mV}}{9 \text{ mA}} = 5.55 \ \Omega.$$

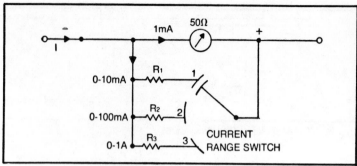

Fig. 14-3. Multirange ammeter.

The total resistance of the instrument is now:

$$\frac{50 \times \dfrac{50}{9}}{50 \times \dfrac{50}{9}} = 5\ \Omega.$$

When the current to be measured is 5 mA, the voltage across the instrument is $5\,\text{mA} \times 5\,\Omega = 25\,\text{mV}$. and the current through the meter movement is $25\,\text{mV}/50\,\Omega = 0.5\,\text{mA}$. The pointer therefore correctly indicates half of the full-scale deflection.

(b). $I = 100\ \text{mA}$, $I_m = 1\ \text{mA}$, $R_m = 50\ \Omega$.

$$R_{sh} = \frac{1\ \text{mA} \times 50\ \Omega}{100\ \text{mA} - 1\ \text{mA}} = 0.505\ \Omega.$$

(c). $I = 1\ \text{A} = 1000\ \text{mA}$

$$R_{sh} = \frac{1\ \text{mA} \times 50\ \Omega}{1000\ \text{mA} - 1\ \text{mA}} = 0.05005\ \Omega.$$

These shunts are normally manufactured from resistance wire with a low temperature coefficient. For currents greater than 20 A, the shunts are too large to be placed within the instrument and are therefore connected externally.

Example 14-3

In Fig. 14-4 an Ayrton shunt is used for the same meter movement and current ranges as in Example 14-2. Calculate the required values for R_1, R_2, R_3.

Solution

The problem is best solved by using the current division rule.
0 – 10 mA Range.

264

$$\frac{R_1 + R_2 + R_3}{50} = \frac{1\ mA}{9\ mA}$$

$$R_1 + R_2 + R_3 = \frac{50}{9}$$

$$R_2 + R_3 = \frac{50}{9} - R_1$$

0 – 100 mA Range.

$$\frac{R_2 + R_3}{R_1 + 50} = \frac{1\ mA}{99\ mA}$$

$$R_2 + R_3 = \frac{R_1 + 50}{99}$$

Therefore:

$$\frac{50}{9} - R_1 = \frac{R_1 + 50}{99}$$

$$550 - 99R_1 = R_1 + 50$$

$$R_1 = 5\ \Omega.$$

and $R_2 + R_3 = \frac{50}{9} - 5 = \frac{5}{9} = 0.55\ \Omega.$

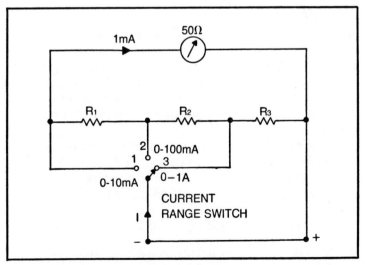

Fig. 14-4. Multirange ammeter with Ayrton shunt.

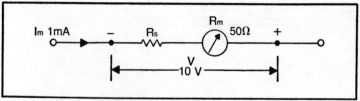

Fig. 14-5. Principle of the voltmeter.

0 − 1 A Range.

$$\frac{R_3}{R_1 + R_2 + 50} = \frac{1}{999}$$

$$999R_3 = R_2 + 55$$
$$999R_3 - R_2 = 55$$

Therefore:

$$1000R_3 = 55\ \frac{5}{9} = 55.55$$

$$R_3 = 0.055\ \Omega$$

Then: $R_2 = 0.55 - 0.055 = 0.5\ \Omega$.
The required values are therefore: $R_1 = 5\ \Omega$, $R_2 = 0.5\ \Omega$, $R_3 = 0.55\ \Omega$.

THE VOLTMETER

As already indicated, the moving coil meter movement is basically a millivoltmeter or microvoltmeter. In order to adapt the movement for higher voltage ranges it is necessary to connect a series multiplier resistor. Most of the voltage to be measured is then dropped across this high value series resistor and only a small part appears across the meter movement to provide the necessary deflection.

In the example of Fig. 14-5 the meter movement once again has a total resistance of 50 Ω and requires a full-scale deflection current of 1 mA. In order to adapt the instrument for the 0 − 10 V range, the total resistance of the voltmeter must be

$$\frac{10\ V}{1\ mA} = 10\,k\Omega$$

and the value of the multiplier resistor is $10\,k\Omega - 50\,\Omega = 9950\ \Omega$. In equation form:

$$R_s = \frac{V}{I_m} - R_m.$$

$$= V \times \text{(sensitivity in } \Omega/V) - R_m$$

where R_s = value of the series multiplier resistor

V = voltage indicated by full-scale deflection.

I_m = full-scale deflection current.

R_m = resistance of meter movement.

In a multirange voltmeter it would be possible to use a separate multiplier resistor for each range. However it is simpler to use a series of resistors as shown in Fig. 14-6.

The moving coil meter movement will not respond to AC directly. For AC measurements, the alternating voltage is usually rectified by a bridge circuit so that the resulting DC can deflect the moving coil meter movement. The instrument may then be calibrated to measure the effective value of the AC voltage.

Example 14-4

In Fig. 14-6 the meter movement has a resistance of 50 Ω and a sensitivity of 20000 Ω/V. The voltage ranges are 0-1 V, 0-10 V, 0-50 V, 0-100 V, 0-500 V. Calculate the required values for the resistors R_1, R_2, R_3, R_4, R_5.

Solution

0 – 1 V Range.

$R_1 = (20000 \ \Omega/V \times 1 \ V) - 50 \ \Omega = 19950 \ \Omega.$

0 – 10 V Range.

$R_1 + R_2 = (20,000 \times 10) - 50 = 199,950 \ \Omega.$

$R_2 = 199,950 - 19,950 = 180,000 \ \Omega.$

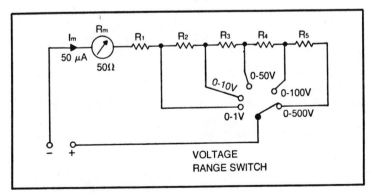

Fig. 14-6. The multirange voltmeter.

0 – 50 V Range.

$R_1 + R_2 + R_3 = (20,000 \times 50) - 50 = 999,950 \ \Omega.$
$R_3 = 999,950 - 199,950 = 800,000 \ \Omega.$

0 – 100 V Range.

$R_1 + R_2 + R_3 + R_4 = (20,000 \times 100) - 50 = 199,950 \ \Omega.$
$R_4 = 1,999,950 - 999,950 = 1,000,000 \ \Omega.$

0 – 500 V Range.

$R_1 + R_2 + R_3 + R_4 + R_5 = 20,000 \times 500 - 50 = 9,999,950 \ \Omega.$
$R_5 = 9,999,950 - 1,999,500 = 8,000,000 \ \Omega.$

THE LOADING EFFECT OF A VOLTMETER

The voltmeter is placed across (in parallel with) the component whose DC voltage drop is to be measured. The ideal voltmeter should have infinite resistance so that the instrument does not load the circuit being monitored. However the resistance of a moving coil voltmeter is far from infinite and its value changes with the particular voltage range selected. For example if the meter movement has a sensitivity of 20,000 Ω/V, the instrument's resistance on the 0-10 V range is $20,000 \times 10 \ \Omega = 200 \ k\Omega$, while on the 0-100 V range, the resistance increases to $20,000 \times 100 \ \Omega = 2$ MΩ. The worst loading effect by such a voltmeter will therefore occur in high resistance, low voltage circuits, as shown in Fig. 14-7 A.

The instrument should measure a voltage drop of 6 V between the points X and Y; the voltmeter is therefore switched to the 0-10 V range and then connected across the lower 200 k Ω resistor, (Fig. 14-7B). However the voltmeter's resistance is also 200 k Ω so that the total resistance of the voltmeter and the resistor in parallel is only 200 k Ω/2 = 100 k Ω. Using the voltage division rule, the reading of the voltmeter is

$$12 \ V \times \frac{100 \ k \ \Omega}{200 \ k \ \Omega + 100 \ k \ \Omega} = 4 \ V.$$

The voltmeter loading effect has therefore reduced the voltage between X and Y from 6 V to 4 V.

The Vacuum Tube Voltmeter (VTVM) and the Field Effect Transistor (FET) voltmeter have a high resistances (several megohms) which are independent of the voltage range chosen such voltmeters have a minimum loading effect.

Example 14-5.

In Fig. 14-7 the voltage drop between X and Y is measured by a voltmeter whose sensitivity is 1,000 Ω/V. What is the reading of the voltmeter when switched to the 0-10 V range?

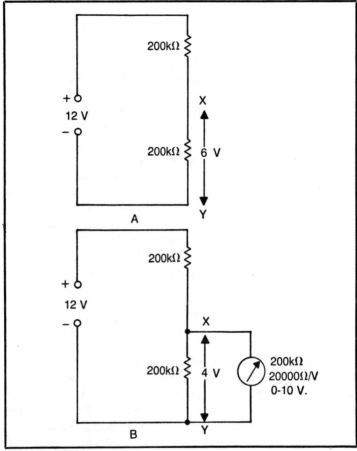

Fig. 14-7. Loading effect of a voltmeter.

Solution

Resistance of voltmeter = 1000 Ω/V × 10 V = 10000 Ω
= 10 k Ω.

Total resistance of the voltmeter and the lower 200 k Ω

resistor in parallel = $\dfrac{10 \times 200}{10 + 200}$ = 9.524 k Ω.

Using the voltage division rule, the voltmeter reading is

$$12\,V \times \frac{9.524 \text{ k } \Omega}{200 \text{ k } \Omega + 9.524 \text{ k } \Omega} = 0.545\ V.$$

THE OHMMETER

The moving coil meter movement may be adapted for resistance measurements by the inclusion of additional resistors and one or more primary cells. Since these cells alone must provide the necessary deflection, it is essential that resistance movements are only made with all power removed from the circuit.

Figure 14-8A shows the circuit of a practical ohmmeter which uses a meter movement with a sensitivity of 20,000 Ω/V and a resistance of 2,000 Ω. The full-scale deflection current will therefore be 50 μA which must flow through the meter movement when the red and black probes are shorted together so that the measured resistance is zero ohms. Shorting the probes together is the initial step in the use of the ohmmeter; the value of the "zero adjust" resistor, R_6, is set to produce the full-scale deflection current which will coincide with the reading of zero ohms on the right hand side of the resistance scale (Fig. 14-8B).

With the range switch in the R × 1 position the value of R_6 is $\frac{1.5 \text{ V}}{50 \text{ } \mu\text{A}} - 1138 \text{ } \Omega - 21850 \text{ } \Omega - 2000 \text{ } \Omega = 30 \text{ k}\Omega - 24988 \text{ } \Omega = 5012 \text{ } \Omega$. The value of R_1 is significant since 11.5 Ω is the value in the center of the scale so that when the unknown resistance, R_x, is 11.5 Ω, the current through the meter movement must be 50 μA/2 = 25 μA. On the left side of the scale the deflection is zero and therefore R_x is an open circuit of infinite ohms. This means that the resistance scale is non-linear as shown in Fig. 14-8 B. The R × 1 range will be used to measure resistances between zero and about 200 Ω although it is difficult to measure accurately on the left hand side of the scale.

When R_x is 11.5 Ω, the cell voltage of 1.5 V will divide equally between R_x and R_1 (ignoring the shunting effect of $R_M + R_2 + R_4 + R_6 = 30$ k Ω across R_1). The voltage across R_1 is therefore 1.5 V/2 = 0.75 V which will drive the necessary current of 0.75 V/30 k Ω = 25 μA through the meter movement.

In the R × 100 range position, the value of R_6 must be reset; R_3 is then included so that when R_x is 11.5 × 100 = 1,150 Ω, the current through the meter movement is 25 μA. This range will be used to measure values of R_x between 200 Ω and 20,000 Ω.

On the R × 10000 range (R_x between 20,000 Ω and 2 M Ω) the additional 6 V cell has to be added in order to drive the required current through the meter movement. The value of R_5 is such that

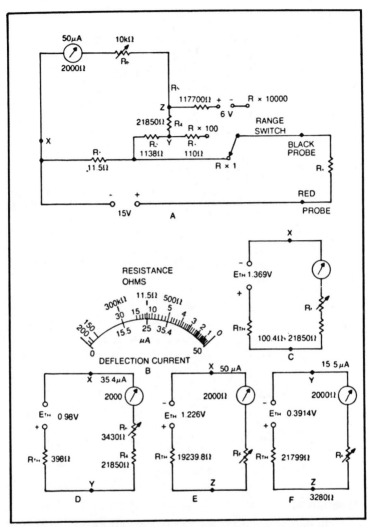

Fig. 14-8. The ohmmeter and its equivalent circuits for examples 14-6 and 14-7.

when R_x is $11.5 \times 10,000 = 115,000\ \Omega$, the deflection current is again $25\ \mu A$.

Example 14-6

In Fig. 14-8A calculate the value for R_6 on the R × 100 range. If R_x is $500\ \Omega$, what is the value of the deflection current through the meter movement?

Solution

With the probes shorted, Thévenize the circuit between the points X and Y (Fig. 14-8C). They by the voltage division rule:

$$E_{TH} = 1.5 \text{ V} \times \frac{(11.5 \ \Omega + 1138 \ \Omega)}{11.5 \ \Omega + 1138 \ \Omega + 110 \ \Omega)}$$

$$= \frac{1.5 \times 1149.5}{1259.5} = 1.369 \text{ V}.$$

$$R_{TH} = \frac{110 \times 1149.5}{110 + 1149.5} = 100.4 \ \Omega.$$

Then: $R_6 = \dfrac{1.369 \text{ V}}{50 \ \mu A} - 21850 \ \Omega - 2000 \ \Omega - 100.4 \ \Omega$

$$= 3430 \ \Omega.$$

When the resistance of 500 Ω is connected between the probes (Fig. 14-8D),

$$E_{TH} = 1.5 \text{ V} \times \frac{(11.5 \ \Omega + 1138 \ \Omega)}{11.5 \ \Omega + 1138 \ \Omega + 110 \ \Omega + 500 \ \Omega}$$

$$= 1.5 \times \frac{1149.5}{1759.5} = 0.98 \text{ V}.$$

and $R_{TH} = \dfrac{610 \times 1149.5}{610 + 1149.5} = 398 \ \Omega.$

The deflection current is:

$$\frac{0.98 \text{ V}}{398 \ \Omega + 3430 \ \Omega + 2000 \ \Omega + 21850 \ \Omega} = 35.4 \ \mu A.$$

Example 14-7.

In Fig. 14-8A, calculate the value required for R_6 on the R × 10000 range. If R_x is 300 k Ω, what is the value of the deflection current through the meter movement?

Solution

With the probes shorted, Thévenize the circuit between the points X and Z (Fig. 14-8E). By the voltage division rule:

$$E_{TH} = 7.5 \text{ V} \times \frac{(11.5 \ \Omega + 1138 \ \Omega + 21850 \ \Omega)}{11.5 \ \Omega + 1{,}138 \ \Omega + 21{,}850 \ \Omega + 117{,}700 \ \Omega}$$

$$= 7.5 \text{ V} \times \frac{22,999.5 \ \Omega}{140,699.5 \ \Omega} = 1.226 \text{ V}.$$

$$R_{TH} = \frac{22,999.5 \times 117,700}{140,699.5} = 19,239.8 \ \Omega.$$

Then: $R_6 = \dfrac{1.226 \text{ V}}{50 \ \mu A} - 19,239.8 \ \Omega - 2,000 \ \Omega = 3,280 \ \Omega.$

When the resistance of 300 k Ω is connected between the probes (Fig. 14-8F),

$$E_{TH} = 7.5 \text{ V} \times \frac{11,999.5 \ \Omega}{440,699.5 \ \Omega} = 0.3914 \text{ V}.$$

$$R_{TH} = \frac{22,999.5 \times 417,700}{440,699.5} = 21,799 \ \Omega.$$

The deflection current is:
$$\frac{0.3914 \text{ V}}{21,799 \ \Omega + 3,280 \ \Omega + 2,000 \ \Omega} = 15.5 \ \mu A.$$

THE WHEATSTONE BRIDGE

This is a precision method of measuring resistances over a wide range of values. The circuit is shown in Fig. 14-9 and consists of four resistance arms, a sensitive current meter (G), and a DC

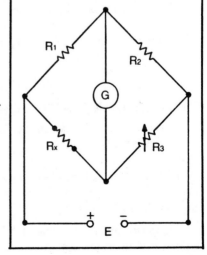

Fig. 14-9. The Wheatstone bridge.

voltage source. R_3 is a accurately calibrated variable resistor while each of the fixed resistors R_1, R_2, only has the possible values of 1 Ω, 10 Ω, 100 Ω or 1,000 Ω. The ratio of $R_1 : R_2$ may then be set for 0.001, 0.01, 0.1, 1, 10, 100 or 1,000.

To measure the unknown resistance, R_x, it is necessary to *balance* the bridge; this is accomplished by selecting the required ratio for $R_1 : R_2$ and then adjusting R_3 until the reading of G is zero.

When the bridge is balanced, the voltage drops across R_2 and R_3 must be equal. Therefore by the voltage division rule:

$$\frac{E \times R_2}{R_1 + R_2} = \frac{E \times R_3}{R_x + R_3}$$

$$R_x = R_3 \times \frac{R_1}{R_2}$$

The measured value of R_m may therefore range from the lowest value of R divided by 1,000 to 1,000 times the highest values of R.

Example 14-8

In Fig. 14-9, $R_1 = 1,000 \Omega$ and $R_2 = 10 \Omega$. When the bridge is balanced, R_3 is 8.2 Ω. What is the value of the unknown resistor, R_x? If the values of R_1 and R_2 are interchanged, what is the new value of R_x?

Solution

Unknown resistance, $R_x = 8.2 \times \dfrac{1,000}{10} = 8,200 \ \Omega = 8.2 \ k \ \Omega.$

When the resistors, R_1, R_2 are interchanged the new value of

$$R_x = 8.2 \times \frac{10}{1,000} = 0.082 \ \Omega.$$

MOVING-IRON AMMETERS AND VOLTMETERS

These instruments can be divided into two types:

1. The attraction type in which a piece of soft iron is attracted towards a solenoid.

2. The repulsion type in which two parallel soft iron vanes are magnetized inside a solenoid and therefore repel each other.

Attraction Moving-Iron Meter

This instrument is illustrated in Fig. 14-10A. The current to be measured flows through the coil and sets up a magnetic flux.

This field magnetizes the soft iron disc and draws it into the center of the coil. The force acting on the iron depends on the flux density and on the coil's magnetic field intensity, both of which are proportional to the current. The torque and the deflection are therefore not proportional to the current alone (as in the moving coil meter movement) but to the square of the current. The result is a non-linear scale which is cramped at the low end but open at the high end. However by careful shaping of the iron disc, the scale's linearity can be improved.

The restoring torque of the controlling device is supplied by a spring or sometimes by gravity. The damping device is commonly obtained by the use of an air piston.

Repulsion Moving-Iron Meter

Two iron vanes are situated axially in a short solenoid as shown in Fig. 14-10B. One vane is fixed while the other is moveable and attached to a pivot which also carries the pointer. When the current flows through the coil, the vanes are equally magnetized and repel each other. The repulsion creates a deflection on the scale which is calibrated for direct reading. The force of repulsion is dependant on the flux density of each iron vane and is directly proportional to the square of the current; therefore the scale is again non-linear. The restoring and damping systems are similar to those of the attraction type.

In the measurement of AC the deflecting torque for both the attraction and repulsion meters will be proportional to the square of the instantaneous current and will therefore vary from zero to a

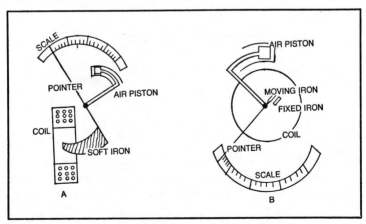

Fig. 14-10. Moving iron instruments.

peak value at a rate equal to twice the alternating frequency. Because of its inertia the moving system will take up a position corresponding to the mean torque. The deflection will then be proportional to the square of the current's rms value.

Unlike the moving coil instrument the moving iron meter may be used for both DC and AC measurements. If the current contains both DC and AC components, the reading will be equal to

$$\sqrt{I_{DC}^2 + I_{rms}^2}.$$

Since the strength of the deflection torque depends on the number of ampere turns for the coil it is possible to arrange for various ranges by winding different numbers of turns on the coil. Also by varying the type of wire used, the meter's resistance can be changed so that there is no need for shunt and swamp resistances.

The moving-iron milliammeter may be converted to a voltmeter by adding a suitable non-inductive series resistor.

Apart from their advantage of being able to measure both DC and AC the moving iron instruments are also relatively cheap and robust but their operation is only satisfactory at line frequencies; moveover they have low sensitivity, are affected by stray magnetic fields and are liable to hysteresis error when used to measure DC.

Example 14-9

A moving iron meter needs 450 ampere turns to provide full-scale deflection. How many turns on the coil will be required for the 0-10 A range? For a voltage scale of 0-300 V with a current of 15 mA, how many turns are required and what is the total resistance of the instrument?

Solution

Number of turns for the 0-10 A range $= \dfrac{450}{10} = 50$ *turns*.

Number of turns for the 0-300 V range $= \dfrac{450}{0.015}$

$$= 30000 \ turns.$$

Total resistance $= \dfrac{300 \text{ V}}{15 \text{ mA}} = 20 \, k \, \Omega.$

THE THERMOCOUPLE METER

The principle of this type of meter is based on the Seebeck effect which was originally discovered in 1806. If a circuit consisting of different metals is at the same temperature throughout, there is no resultant EMF. However if a junction between two

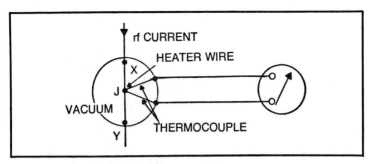

Fig. 14-11. The thermocouple meter.

dissimilar metals is maintained at a different temperature from the rest of the circuit, a *thermoelectric EMF* is generated and a current can then be driven through a conventional moving coil meter movement.

The construction of the meter is shown in Fig. 14-11. The current to be measured passes between X and Y, raising the temperature of the heater wire. Attached to the center of this wire is the thermocouple junction, J; bismuth and antimony are commonly used as the dissimilar metals although many other combinations are possible. As the temperature rises, the thermo-electric EMF will increase and drive a greater current through the meter movement. The amount of deflection on the scale depends on a heating effect which is proportional to the square of the measured current. The meter scale is therefore non-linear so that it is cramped at the low end and open at the high end.

This type of "current-squared" meter is suitable for reading both DC and the rms value of AC. Its particular importance lies in the measurement of radio frequency currents such as occur in the antenna systems of transmitters (Fig. 14-12). Once this type of meter has been calibrated, the calibration is accurate from DC up to microwave frequencies.

Example 14-10

An rf current of 10 A(rms) provides full-scale deflection of a thermocouple ammeter. What percentage of full-scale deflection corresponds to a current of 4 A(rms)?

Solution

The amount of deflection is proportional to the square of the current. Therefore the deflection corresponding to 4 A is 100 × $\left(\dfrac{4 \text{ A}}{10 \text{ A}}\right)$ = *16%* of the deflection.

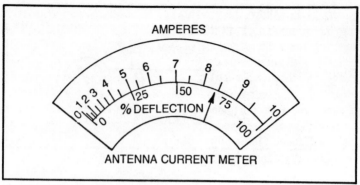

Fig. 14-12. The current-squared meter.

Example 14-11

An rf current of 5 A(rms) corresponds to the full-scale deflection on a thermocouple ammeter. What value of current will produce 60% of the full-scale deflection?

Solution

The current is proportional to the square root of the deflection. Therefore the current corresponding to 60% deflection is 5 A

$$\times \sqrt{\frac{60}{100}} = 3.87 \, A.$$

ELECTRO-DYNAMOMETER MOVEMENT—THE WATTMETER

In the moving-coil meter movement the flux associated with the permanent magnet is fixed in direction; consequently this type of meter (unless used in conjunction with a rectifier) can only be used for DC measurements. However if the permanent magnet is replaced by an electromagnet the direction of the flux may be reversed. This is shown in the *electro-dynamometer* movement of Fig. 14-13A in which two fixed coils F, F' provide the electromagnet. The moving coil, M, is then carried by a spindle and the controlling torque is exerted by spiral hairsprings H, H' which also serve to lead the current into and out of the moving coil.

For DC measurements the electro-dynamometer movement has no advantage over the D'Arsonval type. Compared with AC moving-iron instruments, dynamometer ammeters and voltmeters are less sensitive and more expensive, so they are rarely used. However, dynamometer wattmeters are important since they are the most common way of measuring power directly in AC circuits.

Figure 14-13B shows the way in which the wattmeter is connected into the circuit. The fixed coils are joined in series with the load so that they carry the instantaneous load current. The moving coil is in series with a high value non-inductive multiplier resistor, so that the current through the moving coil is proportional to and nearly in phase with the source voltage. The instantaneous torque on the moving coil is proportional to the product of the instantaneous current through the fixed (current) coils and the instantaneous current through the moving (voltage) coil.

Therefore: Instantaneous torque α instantaneous load current
                                    × instantaneous source voltage
                        α instantaneous power taken by the load.

The controlling torque provided by the hairsprings depends directly on the deflection which is therefore proportional to the power; consequently electro-dynamometer wattmeters have linear scales.

If the load contains both resistance and reactance, the instantaneous power waveform is a second harmonic curve as shown in Fig. 10-1. When the curve is above the zero horizontal line, the torque on the moving coil is clockwise but its direction is reversed when the coil falls below the line. The inertia of the voltage coil and the pointer does not allow the deflection to follow the variations in the instantaneous torque. The rest position of the pointer, P, will therefore represent the average (true) power taken by the load; in other words, the wattmeter automatically takes into account the power factor of the circuit. Therefore: True Power recorded by the wattmeter = E × I × power factor
        = E × I × Cos ϕ.

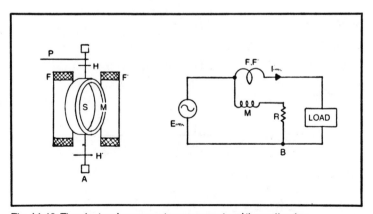

Fig. 14-13. The electro-dynamometer movement and the wattmeter.

where E = rms value of the source voltage

     I = rms value of the load current

    ϕ = phase angle between the source voltage and the load current.

When the load is entirely reactive, the clockwise and counterclockwise torques will exactly balance so that the wattmeter will read zero. However heavy currents may be flowing in the fixed and moving coils so that the instrument must have both voltage and current ratings.

The power factor of the load may be determined by dividing the wattmeter reading by the product of the source voltage and the load current. However, power factor meters have been developed to indicate continuously the value of the power factor and whether it is leading or lagging.

Example 14-12

Readings as taken at an AC source are: voltmeter 110 V, ammeter 6.2 A, wattmeter 590 W. What are the values of the power factor and the phase angle between the source voltage and the inductive load current?

Solution

$$\text{Power Factor} = \frac{\text{True Power}}{\text{Apparent Power}}$$

$$= \frac{590 \text{ W}}{110 \text{ V} \times 6.2 \text{ A}} = 0.865, \text{ lagging.}$$

Phase angle, ϕ = Inv. Cos. 0.865 = $+30.1°$.

CHAPTER SUMMARY

☐ *Moving-coil Meter Movement.*

Torque exerted on coil = BINLd = BANI meter-newtons.

Current, $I = K \times \theta°$ (deflection).

Sensitivity (ohms per volt)

$$= \frac{1}{\text{Full-scale deflection current (amperes)}}$$

☐ *Ammeter*

    $I = I_m + I_{sh}$.

    $I_{sh} \times R_{sh} = I_m \times R_m$.

280

$$R_{sh} = \frac{I_m \times R_m}{I_{sh}} = \frac{I_m \times R_m}{I - I_m}$$

☐ *Voltmeter*

$$R_s = \frac{V}{I_m} - R_m.$$

☐ *Wheatstone Bridge*

$$R_x \times R_2 = R_3 \times R_1$$

$$R_x = \frac{R_1}{R_2} \times R_3.$$

☐ *Current-squared (Thermocouple) Meter*

Meter deflection α (Current)^2

Current $\qquad \alpha \sqrt{\text{Meter Deflection.}}$

☐ *Wattmeter*

True Power in watts = E × I × power factor.

$$\text{Power Factor, p.f.} = \frac{\text{Wattmeter reading}}{E \times I.}$$

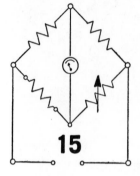

AC Generators-
Polyphase Systems

15

In Chapter 9 the basic alternator contained a revolving armature whose conductors cut the flux of a stationary magnetic field. The alternating voltage generated was then taken through slip-rings to the load. The contacts between these slip-rings and the carbon brushes were subject to friction wear and sparking; moreover they were liable to arc over at high voltages. These problems are overcome in the revolving field alternator of Fig. 15-1.

THE PRACTICAL ALTERNATOR

In this type of generator, a DC source drives a direct current through sliprings, brushes and the windings on a rotor which is driven around by mechanical means. This creates a rotating magnetic field which cuts the conductors embedded in the surrounding stator. An alternating voltage then appears between the ends, S,F, of the stator winding. Since S and F are fixed terminals, there are no sliding contacts and the whole of the stator winding may be continuously insulated.

The alternator of Fig. 15-1 has two poles so that one complete rotation of the rotor generates one cycle of AC in the stator. Therefore, in order to generate 60 Hz, the rotor must turn at 60 revolutions per second or 60 × 60 = 3600 r.p.m. However, in the four pole machine of Fig. 15-2 one revolution would produce two cycles of AC voltage and therefore it would only require a rotor speed of 1800 rpm to generate a 60 Hz output. It follows that:

$$\text{Generated Frequency, } f = \frac{N_p}{60} \text{ Hz.}$$

where: N = rotor speed in r.p.m.

p = number of pairs of poles.

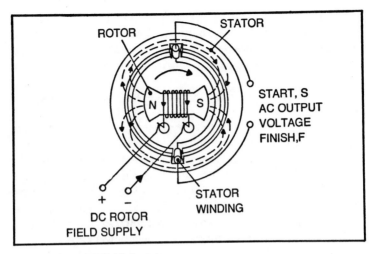

Fig. 15-1. Revolving field alternator.

The power delivered to a resistive load by a single phase alternator is fluctuating at twice the line frequency (chapter nine). This presents a problem since the load on the mechanical source of energy and therefore the necessary torque will not be constant.

Example 15-1

An 8 pole rotor is revolving at 750 r.p.m. What is the value of the generated frequency?

Solution

Generated Frequency, $f = \dfrac{N_p}{60} = \dfrac{750 \times 4}{60} = 50$ Hz.

THE TWO PHASE ALTERNATOR

In this type of alternator two equal coils are mounted on the stator with their axes separated by 90°. Fig. 15-3A shows a

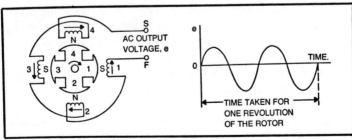

Fig. 15-2. Alternator with four pole rotor.

simplified arrangement of the windings which are cut by the magnetic flux of a two pole rotor. The EMFs e_1, e_2 induced in the windings will therefore be equal in magnitude but 90° out of phase (Fig. 15-3B). These voltage may therefore be represented by $e_1 =$

$$E_{max} \text{ Sin } \omega t \text{ and } e_2 = E_{max} \text{ Sin } (\omega t + \frac{\pi}{2}) = E_{max} \text{ Cos } \omega t,$$

where E_{max} is the peak value of the phase voltage. The fact that e_2 leads e_1 by 90° is a result of the assumed clockwise movement of the rotor. This phase relationship is commonly represented by drawing the two stator windings 90° apart (Fig. 15-2C). Each of the windings may be connected to separate loads but it is more convenient to use a neutral line so that the number of lines required is reduced from four to three as in Fig. 15-3D. Assuming identical (balanced) resistive loads, the instantaneous powers are:

$$P_1 = \frac{e_1^2}{R} = \frac{E_{max}^2 \text{ Sin}^2 \omega t}{R}$$

$$\text{and } p_2 = \frac{e_2^2}{R} = \frac{E_{max}^2 \text{Cos}^2 \omega t}{R}$$

$$\text{Then: } p_1 + p_2 = \frac{E_{max}^2}{R}$$

which is independent of time. The total instantaneous power of the two phase alternator is therefore constant as opposed to the fluctuating power output of the single phase machine. Notice that this advantage is only achieved if the loads are balanced.

If the current in each load is I_{rms}, the neutral line current is

$$\sqrt{I_{rms}^2 + I_{rms}^2} = \sqrt{2} I_{rms} \text{ (Fig. 15-3E)}.$$

The total current in the three lines is therefore $(2 + \sqrt{2}) I_{rms}$. If the same loads were connected in parallel across a single phase alternator, the total current in the two supply lines would be $4 I_{rms}$. A single phase system therefore requires more copper to supply a particular load at a given voltage.

If the two voltages are connected to two pairs of coils whose axes are perpendicular, the two fluxes associated with the coils will combine to produce a rotating magnetic field whose angular velocity is equal to that of the alternating voltage. This principle is used in AC induction and synchronous motors.

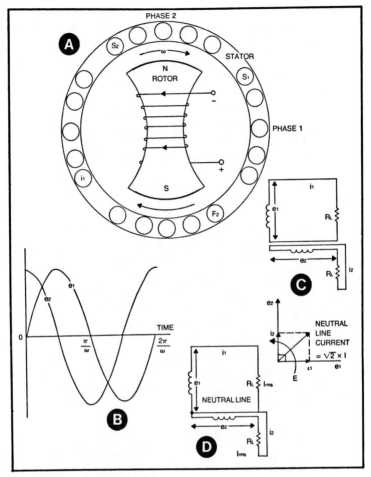

Fig. 15-3. The two phase alternator.

Compared to single phase alternators, high power two and three phase generators require smaller stators, are more efficient and are less subject to vibration.

Example 15-2

A two phase alternator has a 4 pole rotor whose speed is 1600 rpm. What is the value of the generated frequency? The rms voltage of each phase is 120 V and the resistive loads are balanced so that each has an effective resistance of 40 Ω. Find the current in the neutral line and the total instantaneous power output of the generator.

Solution

The generated frequency is independent of the number of phases.

$$\text{Generated Frequency} = \frac{1500 \times 2}{60} = 50 \text{ Hz.}$$

$$\text{Load current for each phase} = \frac{120V}{40\Omega} = 3 \text{ A rms.}$$

$$\text{Neutral line current} = \sqrt{3^2 + 3^2} = 4.242 \text{ A rms.}$$

$$\text{Total instantaneous power} = \frac{E_{max}^{\;2}}{R} = \frac{(120\sqrt{2})^2}{40}$$
$$= 720 \text{ W.}$$

This is also the average power delivered over the cycle.

Note: For the equivalent single phase alternator the total load is equal to 40 Ω/2 = 20Ω. The maximum instantaneous power is

$$\frac{(120\sqrt{2})^2}{20} = 1440 \text{ W,}$$

and the average power over the cycle is 1440 W/2 = 720 W. The total of the currents in the two supply lines is 2 × 120/20 = 12 A; this compares with the two phase alternator where the total of the currents in the two supply lines and the neutral line is (2 × 3) + 4.242 = 10.242 A.

THE THREE PHASE ALTERNATOR

As shown in Fig. 15-4A the three phase alternator has three single-phase windings which are so spaced on the stator that the voltage induced in each winding is 120° out of phase with the voltages in the other two windings (Fig. 15-4B). It should be emphasized that in the three phase alternator the windings are independent of each other and could be connected to separate loads; this would require a six-line system (Fig. 15-4C).

The three voltage waveforms may be represented by their equivalent phasors are shown in Fig. 15-4D.

If V_1 is used as the reference phase voltage and the rotor is assumed to be revolving in the clockwise direction:

$V_1 = E_{max} \text{ Sin } \omega t.$

$V_2 = E_{max} \text{ Sin } \left(\omega t - \frac{2\pi}{3} \right) = - \frac{E_{max} \text{ Sin } \omega t}{2}$

$$-\frac{\sqrt{3}}{2} \; E_{max} \; Cos \; \omega t.$$

$$V_3 = E_{max} \; Sin \left(\omega t - \frac{4\pi}{3} \right) = E_{max} \; Sin \left(\omega t + \frac{2\pi}{3} \right)$$

$$= -\frac{E_{max} \; Sin \; \omega t}{2} + \frac{\sqrt{3}}{2} \; E_{max} \; Cos \; \omega t$$

From these equations it is apparent that:

$$V_1 + V_2 + V_3 = 0.$$

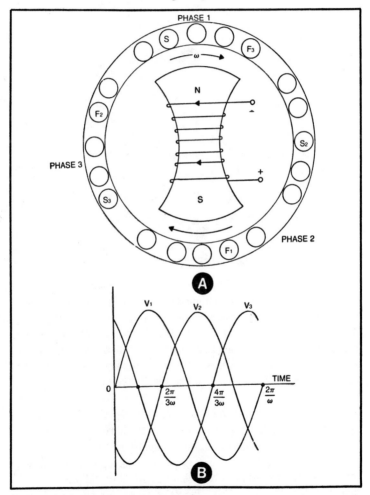

Fig. 15-4 A & B. The three phase alternator.

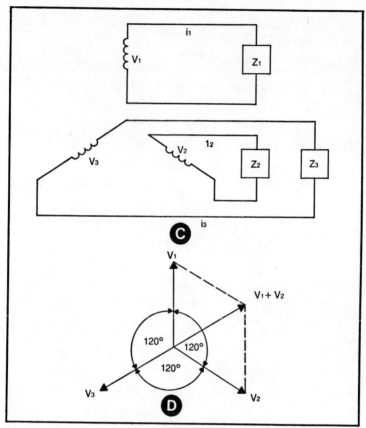

Fig. 15-4 C & D. Six wire three phase connection.

This is also obvious from Fig. 15-4D where the phasor addition $v_1 + v$ is equal in magnitude to v_3 but is exactly opposite in direction. At any point in time the sum of the instantaneous phase voltages is therefore zero. It is then possible to connect three ends of the windings to a common point and the other ends to separate terminals. A single neutral line may be connected to the common point to produce a four-wire Wye (Y) system (chapter eight).

Assume that the three loads are directly connected across the phase windings, (Fig. 15-5A). The loads are balanced in the sense that they all have the same magnitude and power factor (phase angle) which in this case is assumed be lagging. The three load currents will also be equal in magnitude and 120° out of phase with each other (Fig. 15-5B) so that their phasor sum as carried by the neutral line is zero. The neutral line can therefore have a low current capacity since any current which it carries will only exist as a result of imbalance between the loads. Alternatively the neutral line may be replaced by a ground return. The saving in the copper required by the supply lines will therefore be greater for three phase than for two phase (see example 15-3).

If each of the balanced loads is equal to a resistance, R, the instantaneous powers delivered by the phases are:

$$p_1 = \frac{E_{max}^2}{R} \; Sin^2 \; \omega t$$

$$p_2 = \frac{E_{max}^2}{R} \; Sin^2 \left(\omega t - \frac{2\pi}{3} \right) = \frac{E_{max}^2}{R}$$

$$\left(\frac{- Sin \; \omega t + \sqrt{3} Cos \; \omega t}{2} \right)$$

$$p_3 = \frac{E_{max}^2}{R} Sin^2 \left(\omega t + \frac{2\pi}{3} \right) = \frac{E_{max}^2}{R}$$

$$\left(\frac{Sin \; \omega t + \sqrt{3} \; Cos \; \omega t}{2} \right)^2$$

Then:

$$p_1 + p_2 + p_3 = \frac{E_{max}^2}{R} \left[Sin^2 \omega t + \frac{Sin^2 \omega t}{4} + \frac{Sin^2 \omega t}{4} \right.$$
$$\left. + \frac{3}{4} \; Cos^2 \omega t + \frac{3}{4} \; Cos^2 \omega t \right]$$

$$= \frac{3}{2} \times \frac{E_{max}^2}{R} = \frac{3 E_{rms}^2}{R}.$$

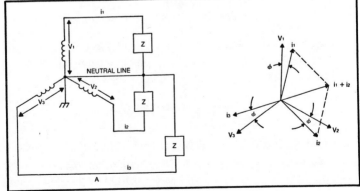

Fig. 15-5. The four wire wye system.

The total instantaneous power for a balanced three system is therefore constant and independent of time.

Example 15-3

A three phase alternator has a 4 pole rotor whose speed is 1800 rpm. What is the value of the generated frequency? The rms value of each phase is 120 V and the resistive loads are balanced so that each has an effective resistance of 60 Ω. Find the total of the line and neutral currents, and the total instantaneous power output of the alternator.

Solution

$$\text{Generated Frequency} = \frac{1800 \times 2}{60} = 60 \text{ Hz.}$$

$$\text{Load current for each phase} = \frac{120 \text{ V}}{60} = 2 \text{ A.}$$

Neutral line current is zero for a balanced load system.
Total of the load and neutral currents = $3 \times 2A = 6A$.
Total instantaneous power = Average power over the cycle

$$= 3 \times \frac{(120 \text{ V})^2}{60} = 720 \text{ W.}$$

Note: For the equivalent two phase alternator the load on each phase would be $2 \times 60/3 = 40$ Ω. The supply line current for each phase is then 120 V/40 Ω = 3 A and the neutral line current is $3 \times \sqrt{2} = 4.242$ A. The total of the line and neutral currents is $(2 \times 3) + 4.242 = 10.242$ A.

Total instantaneous power = Average power over the cycle

$$= 2 \times \frac{(120 \text{ V})^2}{40 \text{ Ω}} = 720 \text{ W.}$$

With the single phase alternator the total load would be $60/3 = 20$ Ω. The total current in the two supply lines is then 2×120 V/20 Ω = 12 A and the average power over the cycle is

$$\frac{(120 \text{ V})^2}{20 \text{ Ω}} = 720 \text{ W.}$$

THE LINE VOLTAGE OF THE WYE SYSTEM

As explained earlier, the EMF applied to the balanced loads was the voltage which existed between each of the lines and ground (or the neutral line). This EMF is normally referred to as the phase

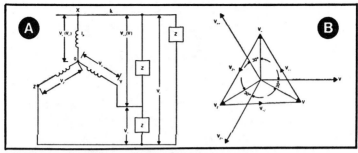

Fig. 15-6. The line voltage of the wye system.

voltage, V_p. However, with three phase systems it is normal to make use of the line voltage as well as the phase voltage. As shown in Fig. 15-6A, the line voltage, V_L, is the EMF existing between any two of the lines. One of the line voltages therefore exists between points X, Y and this voltage will be related to the EMFs generated in the first and second phases of the stator windings.

In determining the relationship between V_L and V_p, it is important to realize that the correct stator terminals must be connected to the common point if the phasor voltages are to be represented as 120° apart; for example, if the two terminals of one winding are reversed, its phase voltage will be shifted by 180°. The phase voltages V_X, V_Y, V_Z are the AC voltages monitored at the points X, Y, Z with respect to the common point 0. Consequently, the line voltage V_{XY} is the AC voltage at X with respect to Y and will therefore be the voltage difference between the two points. In other words $V_{XY} = V_X - V_Y$. The phasor line voltages are:

$$V_{XY} = V_X - V_Y$$
$$V_{YZ} = V_Y - V_Z$$
$$V_{ZX} = V_Z - V_X.$$

If: V_X $E \underline{/0}$, $V_Y = E$ $\underline{/-2\pi/3}$, $V_Z = E \underline{/+2\pi/3}$

$V_{XY} = E\underline{/0} - E \underline{/-2\pi/3}$

$= E - E \left[\left(-\dfrac{1}{2} \right) + j\left(-\sqrt{3}/2 \right) \right]$

$= E \ (3/2 + j\sqrt{3}/2) = \sqrt{3} E \underline{/\pi/6}$

$V_{YZ} = E \underline{/-2\pi/3} - E \underline{/+2\pi/3}$

$= E\left[\left(-\dfrac{1}{2} \right) + j\left(-\sqrt{3}/2 \right) \right] - E \left[\left(-\dfrac{1}{2} \right) + j\left(+\sqrt{3}/2 \right) \right]$

$= \sqrt{3E} \ \underline{/-\pi/2}$

$$V_{zx} = E \underline{\big/ + 2\pi/3} - E \underline{\big/ 0}$$
$$= E \left[\left(-\frac{1}{2} \right) + j \,(+ \sqrt{3}/2) \right] - E \underline{\big/ 0}$$
$$= E \,[(- 3/2) + j \,(+ \sqrt{3}/2)]$$
$$= \sqrt{3}E \underline{\big/ 5\pi/6}.$$

These equations all show that $V_L = \sqrt{3}\ V_p$. Like the phase voltages the line voltages are 120° apart and are shifted by 30° from the phase voltages (Fig. 15-6B).

When three balanced loads are connected between the lines, the three line currents will also be 120° apart. Since each line is directly connected to a terminal of a phase winding, the line current, I_L, must be equal to the phase current, I_p.

Example 15-4

In a three phase system one phase voltage is $120 \underline{\big/ 50°}$ volts. Balanced loads, each equal to $15 \underline{\big/ 20°}\ \Omega$, are connected between the three supply lines. Express the line voltages and the phase currents in their polar forms.

Solution

The phase voltages are $120 \underline{\big/ 50°}$, $120 \underline{\big/ 170°}$ and $120 \underline{\big/ -70°}$ volts.

The line voltages are $120 \sqrt{3} \underline{\big/ 50 + 30°}$, $120 \sqrt{3} \underline{\big/ 170 + 30°}$, $120 \sqrt{3} \underline{\big/ -70 + 30°}$ or $209 \underline{\big/ 80°}$, $209 \underline{\big/ -160°}$ and $209 \underline{\big/ -40°}$ volts.

The phase currents are equal to the line currents which are $209 \underline{\big/ 80°}/15 \underline{\big/ 20°}$, $209 \underline{\big/ -160°}/15 \underline{\big/ 20°}$, $209 \underline{\big/ -40°}/15 \underline{\big/ 20°}$ or $13.9 \underline{\big/ 60°}$, $13.9 \underline{\big/ 180°}$, $13.9 \underline{\big/ -60°}$ amperes.

THE DELTA (Δ) CONNECTION

Since the sum of the three instantaneous phase voltages is at all times zero, it is possible to connect the phase windings in a delta formation (chapter 8). This is shown in Fig. 15-7A and provided the correct connections are made, there will be zero circulating current in the delta loop. Three supply lines may then be connected to the corners of the delta formation and these will be joined to three loads, also arranged in delta. It is clear that the voltage applied to one of the three loads must be the same as the voltage generated in the corresponding winding across which the load is connected. Therefore in the delta system: Line voltage, V_L = Phase voltage, V_p.

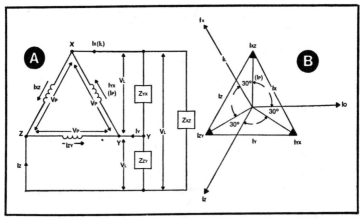

Fig. 15-7. The three phase delta arrangement.

By contrast a particular line current, I_X is associated with two phase currents I_{YX} and I_{XZ}. Each of the load currents will then be equal to its corresponding phase current. However, each line current will be equal to the phasor difference between two of the load currents, so that:

$$I_X = I_{XZ} - I_{YX}$$
$$I_Y = I_{YX} - I_{ZY}$$
$$I_Z = I_{ZY} - I_{XZ}.$$

The sum of the three line currents is therefore zero; this is true irrespective of whether the loads are balanced or not. However, if the loads are balanced, the phase currents will be 120° apart; the relationships between the line and phase currents of the delta system will then be comparable to the equations relating to the line and phase voltages in the wye system (Fig. 15-7B). Therefore, in the delta arrangement:

Line current. $I_L = \sqrt{3} \times$ Phase current, I_p and each line current is shifted by 30° from one of the phase currents.

In balanced wye or delta system the power in each load will be the same and therefore the total power is:

$$P_T = 3 \times E_{rms} \text{ (load)} \times I_{rms} \text{ (load)} \times \text{Cos } \phi$$

where: Cos ϕ is the power factor of each load. For the wye system:

$$E_{rms} \text{(line)} = \sqrt{3} \, E_{rms} \text{(load)}$$

and I_{rms} (line) = I_{rms} (load)

Then: $P_T = 3 \times \dfrac{E_{rms}(line)}{\sqrt{3}} \times I_{rms}(line) \times \text{Cos } \phi.$

$\quad = \sqrt{3} \times E_{rms}(line) \times I_{rms}(line) \times \text{Cos } \phi \text{ watts}.$

For the delta system:

$\quad E_{rms}(line) = E_{rms}(load)$

and $\quad I_{rms}(line) = \sqrt{3}\, I_{rms}(load)$

Then: $P_T = 3 \times E_{rms}(line) \times \dfrac{I_{rms}(line)}{\sqrt{3}} \times \text{Cos } \phi$

$\quad = \sqrt{3} \times E_{rms}(line) \times I_{rms}(line) \times \text{Cos } \phi \text{ watts}.$

The total power can therefore be measured in terms of the line values and the expressions are the same for both wye and delta systems.

Example 15-5

In a three plase delta-connected alternator the phase voltages are 110 $\underline{/0°}$ V, 110 $\underline{/120°}$ V and 110 $\underline{/-120°}$ V. Each of the three balanced loads is equal to 5.5 $\underline{/25°}$ Ω. Calculate the values of the line currents and the total power delivered from the alternator.

Solution

The phase current are: 110 $\underline{/0°}$/ 5.5 $\underline{/25°}$ = 20 $\underline{/-25°}$ A, 110 $\underline{/120°}$/5.5 $\underline{/25°}$ = 20 $\underline{/95°}$ A and 110 $\underline{/-120°}$/5.5 $\underline{/25°}$ = 20 $\underline{/-145°}$ A.

The line currents are:

$\quad 20 \underline{/-25°} - 20 \underline{/-145°}$
$\quad = 18.1 - j\,8.45 + 16.4 + j\,11.47$
$\quad = 34.5 + j\,3.02 = 34.6 \underline{/5°}\,A.$
$\quad 20 \underline{/-145°} - 20 \underline{/95°}$
$\quad = -16.4 - j\,11.47 + 1.47 - j\,19.9$
$\quad = -14.66 - j\,31.37 = 34.6 \underline{/-115°}A.$
$\quad$ and $20 \underline{/95°} - 20 \underline{/-25°}$
$\quad\quad = -1.74 + j\,19.9 - 18.1 + j\,8.45$
$\quad\quad = -19.84 + j\,28.35 = 34.6 \underline{/125°}A.$

The sum of the line currents is: $(34.5 + j\,3.02) + (-14.66 - j\,31.37) + (-19.84 + j\,28.35) = 0.$

Each of the line currents has a magnitude of $20\sqrt{3} = 34.6$A and each is shifted by 30° from one of the phase currents.

The total power delivered from the alternator is $\sqrt{3} \times 34.6 \times 110 \times \cos 20° = $ *6194 W*.

Example 15-6

In a three phase delta connected alternator the phase voltages are 160 $\underline{/80°}$V, 160 $\underline{/-160°}$V and 160 $\underline{/-40°}$V. The corresponding three unbalanced loads are 20 $\underline{/40°}$ Ω, 40 $\underline{/-30°}$ Ω and 80 $\underline{/130°}$ Ω. Calculate the values of the line currents.

Solution

The phase currents are:

$$\frac{160 \underline{/80°}V}{20 \underline{/40°} \ \Omega} = 8 \ \underline{/60°}A.$$

$$\frac{160 \underline{/-160°}V}{40 \underline{/-30°} \ \Omega} = 4 \ \underline{/-130°}A.$$

$$\frac{160 \underline{/-40°}V}{80 \underline{/130°} \ \Omega} = 2 \ \underline{/-170°}A.$$

The corresponding line currents are:

$8 \ \underline{/60°} - 2 \ \underline{/-170°} = (4 + j\,6.92) - (-1.97 - j\,0.34)$
$\qquad\qquad = 5.97 + j\,7.27$
$\qquad\qquad = $ *9.41 $\underline{/50.6°}$A.*

$2 \ \underline{/-170°} - 4 \ \underline{/-130°} = (-1.97 - j\,0.347) - (2.57 - j\,3.064)$
$\qquad\qquad = 0.60 + j\,2.72$
$\qquad\qquad = $ *2.79 $\underline{/77.6°}$A.*

and $4 \ \underline{/-130°} - 8 \ \underline{/60°} = (-2.57 - j\,3.064) - (4 + j\,6.92)$
$\qquad\qquad = -6.57 + j\,9.99$
$\qquad\qquad = $ *11.96 $\underline{/-123.3°}$A.*

Although the loads are unbalanced, the totals of the line currents are:

$(5.97 + j\,7.27) + (0.60 + j\,2.72) + (-6.57 - j\,9.99) = 0.$

CHAPTER SUMMARY

Generated Frequency, $F = \dfrac{Np}{60}$ Hz.

☐ *Two Phase Alternator with Balanced Loads.*

Total instantaneous power = $\dfrac{E_{max}^2}{R}$ watts.

Neutral line current = $\sqrt{2} \times I_{rms}$ (load).

☐ *Three Phase Alternator.*

Sum of the instantaneous phase voltages is zero.

☐ *Wye System* with balanced loads:

Neutral line current is zero.

Total instantaneous power = $\dfrac{3 E_{rms}^2 \text{ (phase)}}{R}$ watts.

Line voltages are each $\sqrt{3} \times$ phase voltage and are separated from the phase voltages by 30°.

Line current = Phase current.

Total power, $P_T = \sqrt{3} \times E_{rms}$ (line) $\times I_{rms}$ (line) $\times$ Cos ϕ watts.

☐ *Delta System*:

Phasor sum of the line currents is zero.

Balanced loads:

Line currents are each $\sqrt{3} \times$ phase current and are separated from the phase voltages by 30°.

Total power, $P_T = \sqrt{3} \times E_{rms}$ (line $\times I_{rms}$ (line) $\times$ Cos ϕ watts.

Appendix A
The Decibel

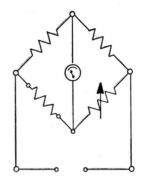

On many occasions electronics is concerned with the transmission of alternating power from one position to another. The various lines and items of equipment that constitute the system introduce gains and losses of power.

POWER RATIOS

Consider a network connecting an alternator to a load. Let the input power be P_i and the output power be P_o. The ratio of output power to input power is then

$$\frac{P_o}{P_i}$$

The network may introduce a loss, in which case

$$\frac{P_o}{P_i}$$

will be less than unity, or it may introduce a gain (for example, an amplifier), in which case

$$\frac{P_o}{P_i}$$

will be greater than unity.

If a number of such networks are connected in cascade (Fig. A-2) and the individual power ratios are known., the overall power ratio

$$\frac{P_o}{P_i}$$

is obtained by multiplying together the individual power ratios. This follows from the fact that:

$$\frac{P_o}{P_i} = \frac{P_{o1}}{P_i} \times \frac{P_{o2}}{P_{o1}} \times \frac{P_{o3}}{P_{02}} \times \ldots \times \frac{P_{03}}{P_{o(n-2)}}$$

or $M_T = M_1 \times M_2 \times M_3 \times \ldots \ldots \times M_{n-1}$

where M_1, etc., are the individual power ratios and M_T is the overall power ratio.

Example A1

In Fig. A-3 calculate the output power at Y if the input power at X is 1 mW.

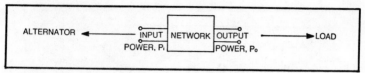

Fig. A-1. Power ratio.

Solution

Overall power ratio, $M_T = 0.215 \times 20.3 \times 0.0246 \times 25.2$

$= 0.679$ (a loss)

Therefore if 1 mW is applied to X, the output power at Y will be *0.679 mW*.

THE LOGARITHMIC UNIT

In a complex system containing a variety of circuits each contributing a gain or loss, calculation of the overall power ratio may become extremely laborious. To simplify this calculation, the individual power ratios are expressed by a logrithmic unit, enabling addition to be employed in place of multiplication. The logarithmic unit employed is the "decibel" (abbreviated to dB), and the power gain or loss, N, of a netowrk expressed in this unit is defined as:

$$N = 10 \log_{10} \frac{P_0}{P_1}$$

where $P_0 = $ the output power

$P_i = $ the input power.

If $\frac{P_0}{P_i}$ is less than unity, then $10 \log_{10} \frac{P_0}{P_i}$ will be negative. A negative sign therefore indicates a power loss, and a positive sign a gain.

It should be noted that, since:

$$10 \log_{10} \frac{P_0}{P_i} = -10 \log_{10} \frac{P_i}{P_0}$$

the numerical answer will be the same no matter whether $\frac{P_0}{P_i}$ or $\frac{P_i}{P_0}$ is considered, but to obtain the correct sign, $\frac{P_0}{P_i}$ must be used.

A larger logarithmic unit is the Bel which is equivalent to 10 decibels and is named after Alexander Graham Bell. The decibel was originally related to acoustics and was regarded as the smallest change in sound intensity which could be detected by the human ear.

The following examples will illustrate the use of the calculator in converting from power ratios to dB and vice-versa.

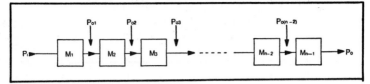

Fig. A-2. Networks in cascade.

Example A2

(1) Convert power ratios of (a) 273 and (b) 0.0469 into decibels.

(2) Convert (a) + 37 dB and (b) − 14.6 dB into their corresponding power ratios.

Solution

(1) (a) For the power ratio of 273, $N = 10 \log_{10} 273$ dB. On the calculator enter "273", then press the "common" or "logarithm to the base 10" key (normally shown as "log"). The display will indicate 2.436 which must finally be multiplied by 10 to give the required answer of *24.36 dB*.

(b) For a power ratio of 0.469 which represents a loss, $N = 10 \log_{10} 0.0469 = -13.3 \, dB$.

(2) (a) To convert + 37 dB into its corresponding power ratio, enter + 37 and then divide by 10. The display indicates 3.7 after which the inverse logarithm key must be used. This key may be shown as Inv. log, $\log^{-3}$, Antilog or 10^x and when pressed, the required power ratio is displayed as *5012*.

(b) Power ratio = Inv. log. $\dfrac{(-14.6)}{10}$ = Inv. log. $(-1.46) = 0.0347$.

Example A3

In Example A1, express each of the individual power ratios as well as the overall power ratio in terms of dB.

SOLUTION

$N_1 = 10 \log M_1 = 10 \log 0.215 = -6.68$ dB.
$N_2 = 10 \log M_2 = 10 \log 20.3 = +13.07$ dB.
$N_3 = 10 \log M_3 = 10 \log 0.246 = -16.09$ dB.
$N_4 = 10 \log M_4 = 10 \log 0.251 = -6.00$ dB.
$N_5 = 10 \log M_5 = 10 \log 25.2 = +14.01$ dB.
$N_T = 10 \log M_T = 10 \log 0.679 = -1.68$ dB.

Then: $N_T = N_1 + N_2 + N_3 + N_4 + N_5$
$= -6.68 + 13.07 - 16.09 - 6.00 + 14.01$
$= -1.69$ dB.

The overall gain or loss in dB for a number of cascaded stages is therefore the algebraic sum of the decibel gains and losses in the individual stages.

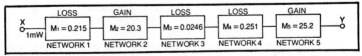

Fig. A-3. Circuit for example A-1.

EXPRESSION OF ABSOLUTE POWER LEVELS IN THE DECIBEL NOTATION

The decibel is fundamentally a unit of power ratio and not of absolute power; but if some standard reference level of power be assumed, then any absolute power can be expressed as so many dB above or below this reference standard. While various other standards may occasionally be encountered, the standard most commonly adopted is 1 mW (0.001 W).

Using this standard, any power P can be expressed as $10 \log_{10} \dfrac{P}{1\,mW}$ dB

referred to 1 mW. Therefore, 1 watt is $10 \log_{10} \left(\dfrac{1000}{1} \right) = 30$ dB above

the standard 1 mW, or "+ 30 dB with respect to 1 mW". Similarly, $5\,\mu W$

is $10 \log_{10} \left(\dfrac{5}{1000} \right) = -23$ dB with respect to 1 mW (23 dB below 1

mW). The expression "dB with respect to 1 mW" or "dB referred to 1 mW" is usually abbreviated to "dBm". Other abbreviations sometimes met are "dB wrt 1 mW", "dB ref 1 mW", and "VU" (volume or voice unit). It therefore appears that the dBm and the VU are the same; however the use of the dBm is normally confined to single frequencies (tones) while the VU is reserved for complex audio signals such as speech or music. However both the dBm and the VU measure audio power levels which are transmitted along a 600 Ω line.

Example A4

(1) Express power levels of (a) 6.38 W and (b) 26.7 μW in dBm.

(2) What power levels are represented by (a) 617 dBm and (b) - 8.6 VU?

Solution

(1) (a) Convert the power level to mW so that 6.38 W = 6.38 × 1000

mW = 6380 mW. Then $6.38\,W = 10 \log_{10} \left(\dfrac{6380\,mW}{1\,mW} \right) = 38.05\,dBm.$

(b) $26.7\,\mu W = 10 \log_{10} \left(\dfrac{26.7/1000\,mW}{1\,mW} \right)$

$\qquad\qquad = 10 \log 0.0267$

$\qquad\qquad = -15.73\,dBm.$

(2) (a) Power level = Inv. log. $\left(\dfrac{+17}{10} \right) = 50.12\,mW.$

(b) Power level = Inv. log $\left(\dfrac{-8.6}{10} \right) = 0.138\,mW.$

$\qquad\qquad\qquad\qquad\qquad = 138\,\mu W.$

CURRENT AND VOLTAGE RATIOS

When it is desired to compare the powers developed in equal input and output resistances, it is sufficient to measure their associated voltages or currents. The power ratio in decibels is then equal to twenty times the logarithm (to base 10) of the current or voltage ratio.

Consider equal input and output resistances of R ohms, which carry currents of rms values I_i and I_o and develop voltages or rms values E_i and E_o respectively. The powers associated with the two resistances are:

$$\text{Input power, } P_i = E_i I_i = R\, I_i^2 = \frac{1}{R} \times E_i^2$$

$$\text{Output power, } P_o = E_o I_o = R\, I_o^2 = \frac{1}{R}\, E_o^2$$

The ratio between these two powers is therefore:

$$\frac{P_o}{P_i} = \frac{E_o I_o}{E_i I_i} = \left(\frac{I_o}{I_i} \right)^2 = \left(\frac{E_o}{E_i} \right)^2$$

or, expressing this in decibels,

$$N = 10 \log_{10} \frac{P_o}{P_i} = 10 \log_{10} \left(\frac{I_o}{I_i} \right)^2 = 20 \log_{10} \frac{I_o}{I_i}$$

$$\text{and } N = 10 \log_{10} \frac{P_o}{P_i} = 10 \log_{10} \left(\frac{E_o}{E_i} \right)^2 = 20 \log_{10} \frac{E_o}{E_i}$$

Therefore the power ratio in decibels is equal to 20 $\log_{10}$ (current ratio) = 20 $\log_{10}$ (voltage ratio), provided that the input and output resistances are equal.

If the input and output powers are associated with input and output impedances, $(Z_i = R_i + jX_i$ and $Z_o = R_o + jX_o)$, the formulas $N = 20 \frac{I_o}{I_i}$ or $20 \log_{10} \frac{I_o}{I_i}$ $20 \log_{10} \frac{E_o}{E_i}$

may still be used for the dB gain or loss provided $R_i = R_o$. If R_i and R_o are not equal, the formulas become:

$$N = 10 \log_{10} \left(\frac{E_o^2/R_o}{E_i^2/R_i} \right) = 20 \log_{10} \frac{E_o}{E_i} + 10 \log_{10} \frac{R_o}{R_i}$$

$$= 20 \log_{10} \frac{E_o}{E_i} - 10 \log_{10} \frac{R_i}{R_o} \, dB.$$

$$\text{and } \quad N = 10 \log_{10} \left(\frac{I_o^2 R_o}{I_i^2 R_i} \right) = 20 \log \frac{I_o}{I_i} + 10 \log_{10} \frac{R_o}{R_i} \, dB.$$

Example A-5

(1). Assuming that the input and output resistances are equal, convert (a) + 43.7 dB and (b) − 27.6 dB into their corresponding voltage (or current) ratios.

(2). Assuming equal input and output resistances, convert into dBs (a) a voltage ratio of 56.3 and (b) a current ratio of 0.00353.

Solution

(1). (a) Voltage (or current) ratio = Inv. log. $\left(\dfrac{43.7}{20} \right)$

$= 153$

(b) Voltage (or current) ratio = Inv. log. $\left(\dfrac{-27.6}{20} \right)$

$= 0.0417$

(2). (a) $N = 20 \log_{10} 56.3 = 35.01\ dB$.
(b) $N = 20 \log_{10} 0.00353 = -49.04\ dB$.

Example A-6

The input resistance of an amplifier is 1.3 kΩ while its output resistance is 17 kΩ. If the amplifier's voltage gain is 35, calculate its power gain in dB.

Solution

Power gain in dB = $20 \log_{10} 35 - 10 \log_{10} \left(\dfrac{17}{1.3} \right)$

$= 30.88 - 8.59$
$= 22.3\ dB$.

EXPRESSION OF ATTENUATION IN DECIBELS AND NEPERS

The decibel is fundamentally a unit of power ratio, but it can be used to express current ratios when the resistive components of the impedances through which the current flows are equal, and voltage ratios when the conductive components of those impedances are equal. The neper is fundamentally a unit of current ratio, but it can be used to express power ratios when the resistive components of the impedances are equal.

The loss of power in an electrical network is know as *attenuation*. Attenuation may be measured using either the decibel or the neper notation.

If the power entering a network is P_i and the power leaving is P_o, then the attenuation in decibels is defined as:

$$\text{Attenuation in decibels} = 10 \log_{10} \frac{P_o}{P_i}$$

If the current entering a network is I_i and the current leaving is I_o, then the attenuation in nepers is defined as:

$$\text{Attenuation in nepers} = \text{Ln} \frac{I_o}{I_i}$$

where Ln is the natural or Napierian logarithm which is based on the exponential $e = 2.7183$

Because of its derivation from the exponential e, the neper is the most convenient unit for expressing attenuation in theoretical work. The decibel, on the other hand, being defined in terms of logarithms to the base 10, is a more convenient unit in practical calculations using the decimal system. The conditions under which the two units may be used can be summarized in the following equations.

$$\text{Attenuation in dB} = 10 \log_{10} \frac{P_o}{P_i}$$

$$= 20 \log_{10} \frac{I_o}{I_i} \text{ provided that } R_i = R_o$$

$$= 20 \log_{10} \frac{E_o}{E_i} \text{ provided that } G_i = G_o$$

$$\text{Attenuation in nepers} = \ln \frac{I_o}{I_i} = \ln \frac{E_o}{E_i} \text{ provided that } Z_i = Z_o$$

$$= \frac{1}{2} \ln \frac{P_o}{P_i} \text{ provided that } R_i = R_o$$

If the resistive components of the impedances at the input and output of the network are equal, then the attenuation may be readily converted from one notation to the other.

$$\text{Attenuation in dB} = 20 \log_{10} \frac{I_o}{I_i}$$

$$= 20 \ln \frac{I_o}{I_i} \times \log_{10} e$$

$$= 8.686 \times \ln \frac{I_o}{I_i}$$

$$= 8.686 \times (\text{attenuation in nepers}).$$

Therefore:

Attenuation in dB = 8.686 × (attenuation in nepers) provided that $R_i = R_o$ or Attenuation in nepers = 0.1151 × (attenuation in dB) provided that $R_i = R_o$.

Example A7

The input current to an electrical network is 27 mA and the output current is 75 μA. Assuming that the input and output resistances are equal, calculate the attenuation of the network in (a) decibels and (b) nepers.

Solution

(a) Current ratio $= \frac{I_o}{I_i} = \frac{75\,\mu A}{27\,mA} = 2.77 \times 10^{-3}$

Attenuation in dBs $= 20 \log_{10}(2.77 \times 10^{-3})$
$= -51.1\,dB.$

(b) Attenuation in nepers $= \ln(2.77 \times 10^{-3})$
$= -5.889\,nepers.$

Check:

$$\frac{\text{Attenuation in dB}}{\text{Attenuation in nepers}} = \frac{51.1}{5.889} = 8.68, \text{ rounded off.}$$

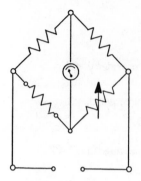

Appendix B
Waveform Analysis

Any single-valued, finite and continuous function, f (t), having a period of 2π, may be expressed in the following form.

FOURIER'S THEOREM

$$f(t) = a_0 + A_1 \sin(\omega t + \phi_1) + A_2 \sin(2\omega t + \phi_2)$$
$$+ A_3 \sin(3\omega t + \phi_3) + \ldots$$

Since $A \sin(\omega t + \phi) = A \sin \omega t \cos \phi + A \cos \omega t \sin \phi$
$$= a \cos \omega t + b \sin \omega t$$
where $a = A \sin \phi$ and $b = A \cos \phi$,
this expansion may be expressed as:

$$f(t) = a_0 + a_1 \cos \omega t + a_2 \cos 2\omega t + a_3 \cos 3\omega t + \ldots$$
$$+ b_1 \sin \omega t + b_2 \sin 2\omega t + b_3 \sin 3\omega t + \ldots$$

In order to use this expression to analyze a complex waveform, it is necessary to determine the coefficients a_0, a_1, a_2, ... and b_1, b_2, ... This is done by multiplying both sides of the above equation by a suitable factor and integrating between the limits 0 and 2π. If the multiplying factor is correctly chosen, all terms vanish except those which will give the required coefficient. This method leads to the following results:

(1) $a_0 = \dfrac{1}{2\pi} \displaystyle\int_0^{2\pi} F(t)\, d(\omega t)$

Note that a_0 is the mean value of f(t) between the limits 0 and 2π.

(2) $a_n = \dfrac{1}{\pi} \times \displaystyle\int_0^{2\pi} f(t) \cos n\omega t\, d(\omega t)$

where n is any positive integer.

Then: a_n is *twice* the mean value of f(t) Cos $n\omega t$ between the limits of 0 and 2π.

(3) $b_n = \dfrac{1}{\pi} \times \displaystyle\int_0^{2\pi} f(t)\ \sin n\omega t\, d(\omega t)$

where n is any positive integer.

Therefore b_n is *twice* the mean value of f(t) Sin nωt between the limits of 0 and 2π.

Example B-1

Find the values of the coefficients in the Fourier analysis of the square waveform shown in Fig. B-1.

Solution

The square wave of Fig. B1 is a single-valued finite and continuous periodic function of ωt, having a period of 2π, and it may therefore be analyzed by Fourier's Theorem.

From t = 0 to t = π, the equation of the function is f(t) = D.

From t = π to t = 2π, the equation of the function is f(t) = 0.

Then:

$$a_0 = \frac{1}{2\pi} \int_0^{2\pi} f(t)\, d(\omega t) = \frac{1}{2\pi} \int_0^{\pi} f(t)\, d(\omega t) + \frac{1}{2\pi} \int_\pi^{2\pi} f(t)\, d(\omega t)$$

$$= \frac{1}{2\pi} \times D \times \pi + 0 = \frac{D}{2}.$$

This verifies that the mean value of f(t) is $\dfrac{D}{2}$.

Also:

$$a_n = \frac{1}{\pi} \int_0^{2\pi} f(t) \cos n\,\omega\, t.\, d\,(\omega\, t)$$

$$= \frac{1}{\pi} \left[\int_0^{\pi} D.\cos n\omega t.d(\omega t) + \int_\pi^{2\pi} 0.\cos n\omega t.d(\omega t) \right]$$

$$= \frac{1}{\pi} \left[\frac{D\sin n\omega t}{n} \right]_0^{\pi} = 0$$

All cosine terms are therefore zero. This is due to the symmetry of the waveform.

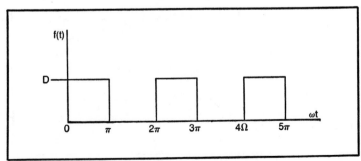

Fig. B-1. The square wave for example B-1.

Finally:

$$b_n = \frac{1}{\pi} \times \int_0^{2\pi} f(t) \, \text{Sin} \, n\omega t \cdot d(\omega t)$$

$$= \frac{1}{\pi} \left[\int_0^{\pi} D \cdot \text{Sin} \, n\omega t \cdot d(\omega t) + \int_{\pi}^{2\pi} 0 \cdot \text{Sin} \, n\omega t \cdot d(\omega t) \right]$$

$$= \frac{1}{\pi} \left[\frac{-D \, \text{Cos} \, n\omega t}{n} \right]_0^{\pi}$$

$$= \frac{D}{n\pi} (1 - \text{Cos} \, n\pi).$$

When n is odd, $(1 - \text{Cos} \, n\pi) = 2$
When n is even, $(1 - \text{Cos} \, n\pi) = 0$

Therefore:

$$b_n = \frac{2D}{\pi} \quad , b_2 = 0 \quad b_3 = \frac{2D}{3\pi} \quad , b_4 = 0, \text{ etc.}$$

The required equation for the square wave is:

$$f(t) = \frac{D}{2} + \frac{2D}{\pi} \left[\text{Sin} \, \omega t + \frac{1}{3} \, \text{Sin} \, 3\omega t + \frac{1}{5} \, \text{Sin} \, 5\omega t + \frac{1}{7} \, \text{Sin} \, 7\omega t + \ldots \right]$$

A waveform frequently encountered in communications is that shown in Fig. B-2. This is similar in general form to the one shown in Fig. B-1, which was represented by the above equation. But it is now symmetrical about the time axis. The first term $\frac{D}{2}$ therefore does not appear in its equation, which is:

$$f(t) = \frac{2D}{\pi} \left[\text{Sin} \, \omega t + \frac{1}{3} \, \text{Sin} \, 3\omega t + \frac{1}{5} \, \text{Sin} \, 5\omega t + \ldots \right]$$

The synthesis of this square wave from its fundamental and harmonic components is shown in Fig. B-3.

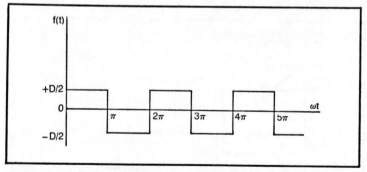

Fig. B-2. Square wave with symmetry about the time axis.

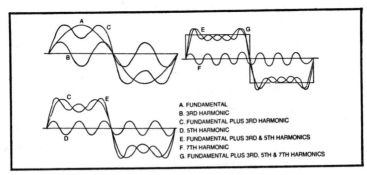

A. FUNDAMENTAL
B. 3RD HARMONIC
C. FUNDAMENTAL PLUS 3RD HARMONIC
D. 5TH HARMONIC
E. FUNDAMENTAL PLUS 3RD & 5TH HARMONICS
F. 7TH HARMONIC
G. FUNDAMENTAL PLUS 3RD, 5TH & 7TH HARMONICS

Fig. B-3. Synthesis of the square wave.

FOURIER'S ANALYSIS OF SOME TYPICAL WAVEFORMS

(1) *Sawtooth (Fig. B-4).*

$$f(t) = \frac{2D}{\pi} \left[\sin \omega t - \frac{1}{2} \sin 2\omega t + \frac{1}{3} \sin 3\omega t - \ldots \right]$$

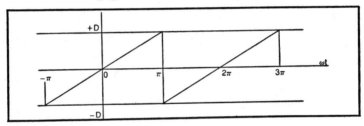

Fig. B-4. The sawtooth wave.

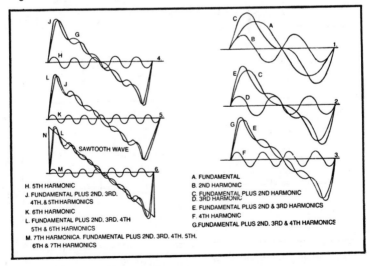

H. 5TH HARMONIC
J. FUNDAMENTAL PLUS 2ND, 3RD,
   4TH, & 5TH HARMONICS
K. 6TH HARMONIC
L. FUNDAMENTAL PLUS 2ND, 3RD, 4TH
   5TH & 6TH HARMONICS
M. 7TH HARMONIC A. FUNDAMENTAL PLUS 2ND, 3RD, 4TH. 5TH.
   6TH & 7TH HARMONICS

A. FUNDAMENTAL
B. 2ND HARMONIC
C. FUNDAMENTAL PLUS 2ND HARMONIC
D. 3RD HARMONIC
E. FUNDAMENTAL PLUS 2ND & 3RD HARMONICS
F. 4TH HARMONIC
G. FUNDAMENTAL PLUS 2ND, 3RD & 4TH HARMONICS

Fig. B-5. Synthesis of the sawtooth wave.

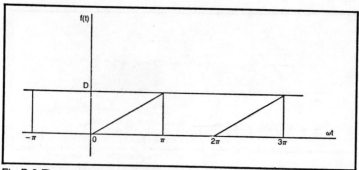

Fig. B-6. The modified sawtooth wave.

The synthesis of the sawtooth wave from its fundamental and harmonic components is shown in Fig. B-5.

(2) *Modified Sawtooth (Fig. B6)*.

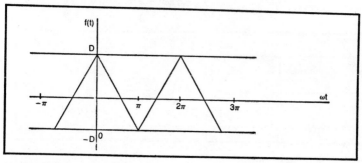

Fig. B-7. The triangular wave.

$$f(t) = \frac{D}{4} - \frac{2D}{\pi^2}\left[\cos\omega t + \frac{\cos 3\omega t}{3^2} + \frac{\cos 5\omega t}{5^2} + \ldots\right]$$

$$+ \frac{D}{\pi}\left[\sin\omega t - \frac{\sin 2\omega t}{2} + \frac{\sin 3\omega t}{3} - \ldots\right]$$

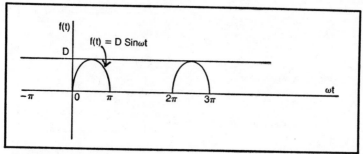

Fig. B-8. The half-wave rectifier output.

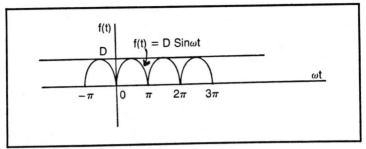

Fig. B-9. The full-wave rectifier output.

(3) *Triangular (Fig. B7).*

$$f(t) = \frac{8D}{\pi^2}\left[\cos\omega t + \frac{\cos 3\omega t}{3^2} + \frac{\cos 5\omega t}{5^2} + \ldots\right]$$

(4) *Half-wave Rectifier Output (Fig. B8).*

$$f(t) = \frac{2D}{\pi}\left[\frac{1}{2} + \frac{\pi}{4}\sin\omega t - \frac{1}{1\times3}\cos 2\omega t - \frac{1}{3\times5}\cos 4\omega t - \frac{1}{5\times7}\cos 6\omega t - \ldots\right]$$

(5) *Full-wave Rectifier Output (Fig. B-9).*

$$f(t) = \frac{4D}{\pi}\left[\ -\frac{1}{1\times3}\cos 2\omega t - \frac{1}{3\times5}\cos 4\omega t - \frac{1}{5\times7}\cos 6\omega t - \ldots\right]$$

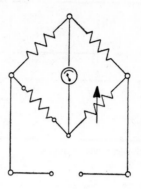

Index

6848 28

Edited by Roland S. Phelps